MASTERING THE COMMODORE 64

Mark Greenshields

This book is dedicated to my parents Jack and Sheila Greenshields.

ACKNOWLEDGEMENTS

With thanks to my parents Jack and Sheila Greenshields, my sister Louise, Douglas Grant, John Lovie, Graeme Douglas, Anita Fabiani and Steve Beats of Commodore. Also special thanks to MICROCOM.

I would also like to thank Tim Hartnell and Richard Gollner for making this book possible.

Mark Greenshields,
February 1983.

First Published in the United Kingdom by Interface Publications

This Remastered Edition
Published by
Acorn Books
www.acornbooks.co.uk

This book is a page-by-page reproduction of the original 1983 edition as published by Interface Publications. The entirety of the book is presented with no changes, corrections nor updates to the original text, images and layout; therefore no guarantee is offered as to the accuracy of the information within.

CONTENTS

1.8 OTHER COMPUTERS' BASICS

This section shows you how to convert other machine's BASICs into C = 64 BASIC. Includes the ZX computer's PRINT AT X,Y ; statement.

1.9 THE PERIPHERALS

This section shows the beginner how to use a printer, a 1540(1) disk drive, joysticks and cassette files on the Commodore 64.

1.10 SPEEDING UP AND IMPROVING YOUR BASIC PROGRAMS

Just as the title says . . .

SECTION 2

2.1 MACHINE LANGUAGE

This section teaches assembly language and machine code on the C = 64.

2.2 COLOUR IN MACHINE CODE

How to use colour in machine code.

2.3 ANIMATION IN MACHINE CODE

This section shows you how to use animation in machine code.

2.4 SOUND AND MUSIC IN MACHINE CODE

This section shows you how to use sound in machine code programs.

2.5 PROGRAMMABLE CHARACTERS IN MACHINE CODE

This section teaches you how to create and use programmable characters in machine code.

2.6 SPRITES IN MACHINE CODE

This section teaches you how to use sprites in machine code.

2.7 COMMODORE 64 ARCHITECTURE AND INTERRUPTS

This section explains how the 64 is put together and how to use the interrupt functions. Includes a machine code program to program the function keys with any thing you want (keywords, etc).

2.8 SIMILARITIES BETWEEN BASIC AND MACHINE CODE

This section shows the similarities between BASIC and machine code and shows how some parts can be directly changed into the other language.

2.9 PROGRAMS

This section contains a Hex loader/saver/enterer, a Hex to decimal conversion program and *vice versa*, and a few miscellaneous programs for you to type in and try.

SECTION 3

THE APPENDICES

This section contains every appendix you would want from a character code table to a complete memory map of the 64 with KERNAL routines listed. It gives the Hex of every address for machine code programmers.

FOREWORD — by Tim Hartnell

The Commodore 64 is one of the most powerful personal computers on the market. It has the graphic capabilities of an arcade machine, the sound of an expensive music synthesiser, and expansion capabilities which make it ideal for many business applications.

But you need a key to unlock the 64s possibilities, and I believe that Mark Greenshields gives you such a key with this exciting book. He takes you through BASIC programming on the Commodore 64, including demonstration programs at every turn to reinforce the information he is sharing, and to make the use of each command and function crystal clear.

From there, you'll progress to assembly language (machine code) programming. By the time you've studied this section, you'll have gained a great deal of competence in this exciting area of programming.

Mark shows you a number of easy ways to control sound, sprites, the high-resolution graphics and colour.

It's time to get underway, so you too can learn to master the Commodore 64.

Tim Hartnell,
March 1983.

Tim Hartnell is author of a number of personal computer books, including 'The Book of Listings' (BBC Publications), 'Getting Acquainted with Your Vic 2Ø' (Interface Publications) and 'The Personal Computer Guide' (Virgin Books).

INTRODUCTION

There are certain expressions which are frequently used in the book. They should be studied and understood before continuing with the book.

VARIABLES

A variable is a letter, series of letters or a letter and a number that stand for a number or a list of letters in a program. This is done to save time and memory and add flexibility. There are three types of variables:

1. A numeric variable.
2. An integer variable.
3. A string variable.

They are all suffixed in different ways to distinguish them from each other. A numeric variable has no suffix:

eg.
a or z or a(1) or a4, etc.

An integer variable is the same as a numeric variable except that it is suffixed with a '%' sign. Their use is primarily to save memory in preference to numeric variables.

eg.
a% or D% or a(1)% or h7%, etc.

A string variable is used to store 'strings' of letters or numbers. This makes text easy to handle. (It is suffixed, with an '$' (dollar sign).

eg.
A$ or f$ or A(1)$ or h7$, etc.

MATHEMATICAL OPERATORS

+ add
− subtract
* multiply
/ divide
↑ to the power of

MASTERING THE COMMODORE 64

The Commodore 64 uses a language called BASIC (Beginners All-purpose Symbolic Instruction Code). It is the simplest of all languages to learn and use. However, it has a few drawbacks. Most of all it is slow. This book will try to explain how to speed up BASIC programs.

MASTERING THE COMMODORE 64

Mark Greenshields

Acorn Books

SECTION 1.1

This section covers the whole of the BASIC language on the Commodore 64 and explains how sound and music synthesis, colour, high-resolution and sprites can be used on the Commodore 64.

The commands are not presented in alphabetical order because it is far easier to learn them in *this* order.

We will begin with the command PRINT.

The command PRINT does exactly what it says. It prints characters on the screen, printer, disk drive or cassette. It has two basic formats:

1) PRINT " "(quote signs). This puts what is inside the quotes onto the screen.
2) PRINT A. This prints the value given to the variable A. If A has not been set then it will PRINT Ø.

When a variable is used after a quote then a semicolon (;) should be used.

eg. PRINT "HELLO YOUR SCORE IS" ; SCORE

Now that we can print variables on the screen we want to change them. This brings us to our next command, LET. The LET command is used to assign a (variable) to a number or string. The variable (numeric) can be assigned a number between:

−1.7Ø141183 E+38 (times ten to the power of 38) to +1.7Ø141183 E+38 for the largest floating point number, and ± 2.93873588 E−39 for the smallest floating point number.

The LET command allows you to add or subtract (or multiply or divide) one number from another.

eg.
```
LET A = A + 1
LET A = A + B + C
```

In every case the LET command can be omitted on the Commodore 64.

eg.
LET A = 1 is the same as A = 1.

So far, as soon as the computer has carried out your command it forgets all about it. To write a program we need a way of making the computer remember our commands. For this we assign *line numbers* to the commands. They are stored and carried out in ascending order. They can be in the range Ø to 63999.

eg.
```
10 PRINT "I COST" ; price "POUNDS"
 5 price = 350
```

Commands with line numbers are referred to as progams. The above program would start with line 5.

Once a program is written we need a way to start it. We do this with the command RUN. This command clears all previously set variables to zero (or strings to empty strings). It then executes the program from the first line number. It can also have a number after it. In the following case the program would start to execute from line 1ØØ.

```
RUN 100
100 PRINT "HELLO"
110 PRINT "GOODBYE"
```

Once you are finished looking at your program you will want to stop it. This is done with the RUN/STOP key (at the left of the keyboard). If you accidentally press this key, you can re-start the program by typing the command CONT. This will only work if you don't change anything in the program.

EDITING

Now that you have written your first program you will want to change certain things. There is one new command and a few utilities which are in the Commodore 64 to help you. The command is LIST. It has four different formats:

1) LIST	Lists the whole program.
2) LIST −100	Lists the program up to and including line 100.
3) LIST 100−	Lists the program from line 100 to the end.
4) LIST 100−200	Lists the program from line 100 to line 200.

The LIST command *puts* the program lines onto the screen for you to see or change.

There is an excellent editing facility on the Commodore 64. There are three special keys (and SHIFT) to help with editing. These are:

1) CRSR (up and down). Move the cursor up or down the screen.
2) CRSR (right and left). Move the cursor right and left.
3) INST DEL. Inserts or deletes spaces into (or characters out) from a program.

When editing you have three options:

1. Re-type the whole line. The new one will replace the old one.
2. Get rid of the line completely by typing the line number and pressing RETURN.
3. Use the cursor and correct the error.

If you are studying a program as it is running then it can be slowed down by pressing the CTRL key.

Now that you have written a program you will probably want to store it for later use. For this you will need the 'C2N Datasette' or the floppy disk drive (see disk handling). There is a command which stores a program on the cassette and this is SAVE. It has two main formats:

1. SAVE. Stores the program on tape without a name.
2. SAVE "NAME". Stores the program on tape and names it 'NAME'.

When you press RETURN after typing this command you will be told to press record and play on tape. Do this and wait (have a tape in, of course). If the computer stops with 'OK. READY' then rewind the tape and type 'VERIFY'. Otherwise rewind the tape and re-SAVE the program.

The command VERIFY checks the program on the tape to see if it is the same as that which is in the computer's memory. If all is OK then it will say so. If not re-SAVE the program.

Once the program is on tape you will want to retrieve it at some time. The command LOAD does just that. It has two basic formats which are just like those of SAVE. They are:

1. LOAD. This loads the first program on the tape, and
2. LOAD "NAME". This loads the program named 'NAME' whether or not it is the first program on the tape. If the program LOADs properly then it will say 'OK'. If you get a load error then try again.

NOTE

Try to keep the tape recorder at least two feet away from the TV. This will help successful SAVing and LOADing. *Keep your cassettes away from the TV!*

It is a good idea to save each program twice on the tape and use computer quality C12 cassettes for more reliable storage.

There is a short cut to typing 'LOAD' to load a program. Press SHIFT and without taking your finger off press RUN/STOP. The program will now automatically load and run.

Often in a program the user is required to enter some information into the computer. There are two commands which deal with this. The first of these is INPUT, which has two formats:

1. INPUT A$ or INPUT A (for strings or numbers respectively).
2. INPUT "ENTER YOUR NAME" ; A$ or INPUT "ENTER YOUR AGE" ; A.

The second format works in exactly the same way as the first except that it PRINTs information before asking for information (with a question mark).

The INPUT command assigns values or strings to the variables just like the LET command only it allows the variable to be changed while a program is running. A skill level in a game is an example.

NOTE

Don't use more than 38 characters in the quotes because if you do the information will also be accepted by the variable giving bad data.

When you have finished with a program you are likely to want to get rid of it from the computer's memory. Instead of switching the computer off you can type the command 'NEW'. It clears the program from memory and sets all variables to zero. Once a program has been NEWed it is lost, unless it has been SAVEd on tape or disk.

If you want to clear the variables but leave the program intact you type 'CLR'.

The other command for inputting information is GET A$. Unlike INPUT it only accepts one letter or number at a time and it cannot print text. Its principal use is where options are given and you are asked to choose A,B,C,D,E or F, for example. In the following program the computer asks for a letter or number. It will then print what you press.

```
10 GETA$
20 PRINTA$
30 RUN
```

SCREEN COMMANDS

The cursor can be moved around the screen, coloured one of 16 colours or reversed with special PRINT commands. To use them you just 'PRINT "' and then the relevant command/s like this.

All the commands are listed below with their meanings.

```
0 PRINT"  s IS CLEAR SCREEN"
10 PRINT" S IS HOME CURSOR"
20 PRINT" r IS BLACK"
30 PRINT" E IS WHITE"
40 PRINT" £ IS RED"
50 PRINT" □ IS CYAN"
60 PRINT" _ IS PURPLE"
70 PRINT" ↑ IS GREEN"
80 PRINT" ← IS BLUE"
90 PRINT" π IS YELLOW"
100 PRINT"a IS ORANGE"
110 PRINT"u IS BROWN"
120 PRINT"v IS LIGHT RED"
130 PRINT"w IS GREY 1"
140 PRINT"x IS GREY 2"
150 PRINT"y IS LIGHT GREEN"
160 PRINT"z IS LIGHT BLUE"
170 PRINT"⊥ IS GREY 3"
```

```
18Ø PRINT"R IS RVS ON"
19Ø PRINT"r IS RVS OFF"
2ØØ PRINT"Q IS CURSOR DOWN"
21Ø PRINT"q IS CURSOR UP"
22Ø PRINT"] IS CURSOR LEFT"
23Ø PRINT"▲ IS CURSOR RIGHT"
```

The colours zero to seven are obtained by pressing CTRL and a key numbered 1 to 7. The colours eight to 15 are obtained by pressing the Commode key (C =) and any number from 1 to 8.

The cursor commands are obtained with the cursor keys at the bottom right of the keyboard.

Reversed characters are turned on by pressing CTRL and 9, and they are turned off with CTRL Ø.

When you are writing a program it is likely that you will want to know how much free memory you have left. The command for this is PRINT FRE (X). This command will only work if you have less than 32K of memory left. So to make this just POKE 52,Ø: POKE 56,136. This gives you seven less K but if you need them then type 52,16Ø: POKE 56,16Ø. The memory is measured in bytes. A byte is one character. So if you have 1Ø24 bytes left you can have 1Ø24 characters more in your program.

The Commodore 64 has an ASCII character set which controls everything from colour and cursor controls to printer commands. They are accessed with the command PRINT CHR$ (ASCII code).

eg.

PRINT CHR$ (147) clears the screen. For a complete list of the CHR$ codes see Appendix 11.

There is a complementary command to CHR$. It returns the value for the character. It is ASC(character). If it is a single character then the format is PRINT ASC("character"). If it is a string then the format is PRINT ASC(A$).

eg.
PRINT ASC("A") gives 65.

STRING HANDLING COMMANDS

A string can be divided up for various purposes. There are three commands that do this:

LEFT$ (string name, number of letters)
RIGHT$ (string name, number of letters)
MID$ (string name, starting position, number of letters).
LEFT$ takes a set number of characters from the left side of the string.

eg.
A$ = "ABCDEFGHIJK"
PRINT LEFT$(A$,2) PRINTs AB

RIGHT$(A$,3) takes the rightmost characters which in this case is IJK.

MID$(A$,3,4) takes characters from the middle starting at position three which in this case is CDEF. These commands are very powerful and allow complex string slicing to take place.

The length of a string can be found with the LEN(A$) statement:

eg.
PRINT LEN (A$)

Which in this case will print 11. The numerical value of a string can be found with VAL(A$). It is only useful for numeric (totally) strings as letters return a zero value. There is a command which returns the numerical value of a string or argument. It is STR$.

eg.
PRINT STR$(12↑22) PRINTS 144

In a program you will often want to know what a certain part of a program does. There is a command which caters for this. When the computer meets this command it ignores what comes after the command and goes immediately to the next line. The command is REM.

eg.

```
10 PRINT CHRS(147)
20 REM LINE 10 CLEARS THE SCREEN
30 REM HELLO *1234567890 ETC.
```

More than one command can be put on one line. They are separated by the colon (:). It is then just as if there was a new line, only it is on the same one.

eg.

```
10 PRINT "HELLO" :REM PRINTS HELLO
```

When a program eventually comes to the end you usually want to start it again. One way is to have the last line as RUN. There is a better way to do this which preserves variables, GOTO line number.

eg.

```
10 PRINT"THE COMMODORE 64"
20 GOTO10
```

There is a command that allows you to go to a *subroutine* or *routine* outside the main program. It is GOSUB. When you have GOSUBed from a program you have to return to it. The command RETURN does just that. It returns to the command after the one you just left.

eg.

```
10 PRINT"S"
20 PRINT"COMMODORE 64"
30 GOSUB60
40 PRINT"COMPUTER"
50 GOTO20
60 PRINT"RπTHE"
70 RETURN
```

It works just like GOTO except that it must be returned from.

A program can be stopped by other methods than pressing the STOP key. There are two commands which do this. The first of these is END. END stops a program when encountered but it cannot be re-started unless RUN or GOTO is used. Its main use is logically to end a program.

The second of these commands is STOP. This command functions in much the same way as END except that the program can be continued by typing CONT. This command is used principally in debugging programs.

The following two programs are the same except that in Program Two, the program can be continued.

```
1 PRINT"THE COMMODORE 64"
2 END
3 GOTO1
```

```
1 PRINT"THE COMMODORE 64"
2 STOP
3 GOTO1
```

In a program there is often a situation where results depend on the value of a number, for example.

eg.
IF a number equals 1Ø THEN do command.

The IF command checks to see if an expression is true. IF it is, THEN the computer does what is after the THEN. If the statement is false THEN the computer goes to the next line.

```
10 A=1
20 IFA=12THENGOSUB100
30 A=A+1
40 PRINTA
```

```
50 GOTO20
100 PRINT"THE STATEMENT IS TRUE"
110 A=1:RETURN
```

Now we have come upon a problem. The program is running too fast to see clearly. This brings us to our next command (well, three).

In the above example we needed to introduce a delay, ie. we wanted the computer to do something for a short time. On the Commodore 64 (or most other computers for that matter) we get it to count from one number to another before continuing. The FOR NEXT loop is what we use. It has the syntax:

'FOR' variable = First number 'TO' second number 'STEP' what to go up or down by.

The STEP tells the computer whether it has to count in ones or halves, for example. The step can be any positive or negative whole or decimal fraction.

The FOR TO STEP loop would only carry out the first part of the statement unless we told it to loop back and do the NEXT one and the NEXT.

eg.

This program gets the computer to count from one to 100 and print it in the top left screen position. As in this program if the step is one (only up) then it can be omitted.

```
10 FORA=1TO100:REM START LOOP
20 PRINT"S";A:REM PRINT NUMBER
30 NEXTA:REM GO BACK AND DO AGAIN UNTIL
A=100
```

The variables can be omitted from the NEXT statement, but it is advisable to leave them in as it makes programs easier to understand. FOR NEXT loops (as they are known) can be nested (one or more inside another) but they must be in the correct order. The innermost loop is completed before any of the rest. In the following example the loops are NEXTed. The 'C' loop completely executes before the 'B' or the 'A' loop. The 'C' loop executes fully for every part of the 'B' loop and 2Ø times for every part of the 'A' loop.

```
10 FORA=0TO20
20 PRINT"A=";A
30 FORB=30TO0STEP-1
40 PRINT"B=";B
50 FORC=200TO250STEP1.5
60 PRINT"C=";C
70 NEXTC,B,A:REM DO C THEN B THEN A
```

Sometimes during a game you want to introduce a random situation into a game, for example. There is a BASIC command which creates a random number. It is RND and it has the syntax: A = RND (1). The one in brackets can be any numeric; it does not alter the value of the random number.

The RND function creates a number between Ø and Ø.99999999. However, these numbers are not always in the range that we want. If we want a number between zero and six we need to multiply the random number by six:

A= RND(1) *6.

But wait. . .we can only get numbers between Ø and 5.99994. We need to add a new command to help us. It is INT. INT returns the integer value of a number (the whole number).

eg.
PRINT INT (5.9999) is five (it just ignores what is after the decimal point).

So to get a random number between one and six we need to use the following:

PRINT INT (RND(1) *6) + 1

The syntax for this is:

A = INT(RND(1)* upper limit + lower limit

The Commodore 64 has two inbuilt time clocks. One which counts in seconds, minutes and hours and one which counts in sixtieths of a second. They are accessed by the statements:

TIME$ is the hours, minutes and seconds clock.
TI is the sixtieth of a second clock.

To reset the clock type: TI$ = "000000". This resets both the TI and TI$ clocks.

Often in a program you will come across a need for numeric or string data. This will need to be read from somewhere (just as you would read your 'data' from a book). The commands that do this are:

READ (a string or numeric variable).
DATA (the actual data to be read).

The following example takes a number from the DATA and multiplies it with A.

```
10 PRINT"s"
20 A=12
30 FORC=0TO12
40 READB
50 PRINTA*B
60 NEXTC
70 DATA 0,1,2,3,4,5,6,7,8,9,10,11,12
```

If the DATA is to be read again it must be freed from the computer. The command is RESTORE. It stands by itself:

```
100 RESTORE
```

If subscripted variables are to be used then space must be allocated to them in the computer's memory. The comand that does this is DIM. It has the syntax:

DIM variable (number of numbers − 1) a string, numeric or subscripted variable.

The number of divisions inside the brackets is limitless, but DIM statements are 'memory greedy' so just because you have 40K (for BASIC) at your disposal you can still run out of memory if you use too large a number!

DIM A (999) is OK and so is DIM A (1,1,1,1,1,1,1,1,1,1) but DIM A (9999) or DIM A (1,1,1,1,1,1,1,1,1,1,1,1,1,1,1,1,1,1) are too big. You can only DIMension an array (list of numbers or letters) once so put these at the beginning of the program. If you re-DIM an array you will get an 'RE DIM'd ARRAY ERROR'.

A subscripted variable is one that has a number in brackets after it.

eg.
A(12) or A(12,12,12), etc

The use of a subscripted variable is mainly in adventure games or in programs with a lot of information which is released depending on the value of one variable or result of a calculation.

In mathematical calculations it is often necessary to evaluate complex equations frequently. They are usually long and memory consuming. Why not create the function and call it as it is needed. To define the function we use the DEF FN command. To call it we use the FN command. The FN of each has a variable after it to tell the computer which function to call.

eg.

```
5 A=12
10 DEFFNX(X)=(A*(12↑2))
20 PRINTFNX(A)
30 A=A*2:GOTO20
```

The syntax is:

DEF FN variable (X) = (X expression)
and A = FN variable (x) or number

The Commodore 64 comes equipped with trigonometric and scientific statements:

SIN (X)
COS (X)
TAN (X)

They all give the value of X where X is an angle measured in radians. A radian is 57.29577 degrees. So, to calculate the value of X when X is in degrees use the following formula:

SIN (degrees/57.2957795)
or COS or TAN.

eg.
FIND the SIN of 30....
PRINT SIN (30/57,2957795)

The answer is .5 which is correct. The same formula applies for COS and TAN as well.

When finding the inverse, ie. going from the value of X to X, you multiply by 57.2957795 to get the degrees. There is only one inverse trigonometric function and that is ATN.

eg.
If the value of X is .5, what is X?
PRINT ATN (.5 * 57.2957795)
Answer 26.5650512

Although the Commodore 64 does not have the inverse SIN and COSINE they can be calculated using the following equations:

INVERSE SINE = ATN(X/SQR(−X*X+1) in radians or
INVERSE SINE = ATN(X/SQR(−X*X+1)*57.2957795 in degrees.

INVERSE COSINE = X − ATN(X/SQR(−X*X+1))+π/2 in radians or

INVERSE COSINE=ZZZZ(-ATN(X/SQR(-X*X+1))+π/2)*57.2957795 in degrees

But now we see the SQR function. What does it mean? Well, it calculates the square root of a number.

eg.
PRINT SQR (81)
Answer 9

The number in the brackets must be greater than or equal to zero.

Often in maths you just want the answer to a question — you don't want to know its sign. There is a BASIC command which deals with this. It is ABS.

eg.
PRINT ABS(−999) gives 999 but PRINT ABS(999) also gives 999.

If you just want to know the sign but not the answer, the statement SGN tells you this.

eg.
PRINT SGN (−999) gives −1 but PRINT SGN (999) gives 1.

The result will be one if the answer is positive, minus one if negative and zero if zero.

The LOG of a number can be calculated with the LOG statement. It is the natural log. To convert to LOG base 1Ø simply divide by LOG(1Ø). To get the inverse of the LOG you use EXP. This takes the log of the number and puts the mathematical constant e (2.71827183) to the power of the LOG.

eg.
PRINT LOG (999)
Answer 6.90675478
PRINT EXP (6.90675478)
Answer 998.79325 which is rounded up to 999.

Instead of using cursor commands (in quotes) you can use a faster and more memory economic method with the two commands, TAB and SPC. These commands are used in conjunction with the PRINT command. The TAB command moves the cursor to the column specified.

eg.
PRINT TAB(12) moves the cursor to column 12.

SPC moves the cursor a number of spaces forward.

eg.
PRINT SPC(12) moves the cursor 12 spaces forward.

Both perform the same function except that when SPC is used after TAB the cursor does not move to the next line but X spaces forward.

The position of the cursor (column) can be found by the command POS.

eg.
PRINT POS (X)
result 22.

The character in the brackets can be any printable one letter or numeral. It only works in program mode or after a cursor command or control character(s).

There are situations in programs where binary comparisons need to be made. A binary number is a number which can be one or zero. In the Commodore 64 or any other computer with an '8 bit' microprocessor, the maximum number is 255 which is 11111111 in binary. In binary a one stands for on and a zero stands for off. To change a decimal number to binary or *vice versa* use the following table. If a one is in the box below a number, then add that number to the total.

128	64	32	16	8	4	2	1		DECIMAL
									BINARY
1	Ø	1	Ø	Ø	1	Ø	1	=	165

The main use of binary comparisons is in machine code where there are no 'greater than' or 'less than' commands. It is also used in character designing. See their respective sections for more information. There are three binary comparison commands in BASIC. They are AND OR and NOT.

They all add two binary numbers together and depending on which command is used the answer appears.

With AND there must be a one in both boxes for a one to be put into the third.

	1	Ø	1	1	Ø	1	1	1	183
AND	1	1	1	1	1	Ø	Ø	Ø	248
	1	Ø	1	1	Ø	Ø	Ø	Ø	176

The other command for binary comparison is OR. It functions like AND except that if a one is in either box then a one is put in the third.

	1	Ø	1	1	Ø	1	1	1	183
OR	1	1	1	1	1	Ø	Ø	Ø	248
	1	1	1	1	1	1	1	1	255

The statement NOT is used where you want the program to do something if the answer is false.

eg.
p = 1

p=p−1: IF NOT P THEN 1ØØ

The program will branch to line 1ØØ if P is not equal to 1. AND, OR and NOT can also be used in IF THEN statements such as:

IF A = 1 AND B = 2 THEN 1ØØ. If and only if A=1 and B=2 then GOTO line 1ØØ.

IF A = 1 OR B = 2 THEN 1ØØ. If A=1 or B=2 then GOTO line 1ØØ.

IF NOT THEN 1ØØ. IF A=Ø then GOTO line 1ØØ.

FILES AND TAPE/DISK/PRINTER USAGE

All these commands will be fully explained in the printer and data file sections. But here is a short summary. The command, OPEN tells the computer to send the data to other devices except the screen. However, it only opens the way. Other commands need to be used to actually move the data. It requires one, two or three operators.

The devices that can be 'OPENED' are:

DEVICE	DEVICE NUMBER
Keyboard	Ø
Cassette	1
RS 232	2
Screen	3
Printer	4
Printer	5 (depending on switch on printer)
Disk drive	8
Serial bus	4−127
Serial bus	128−255 (if after carriage return)

It has the syntax:

OPEN device no, command #, string.

The command ' # ' is used to tell the computer which mode to put the OPEN into.

DEVICE	COMMAND #	EFFECT
Cassette	Ø	Read tape file.
Cassette	1	Write tape file.
Cassette	2	Write tape file with end of tape marker at end of file.
Disk	1–14	Open data channel.
Disk	15	Open command channel.
Keyboard	1–255	No effect.
Screen	1–255	No effect.
Printer	Ø	Upper case/graphics mode.
Printer	7	Upper lower case.

eg.

OPEN 1,Ø	Read the keyboard.
OPEN 1,1,Ø,"name"	Read from cassette.

After a device has been opened and when its use is over, the computer must be told. The CLOSE command does this. It has the same syntax as the OPEN command. So if you OPEN 1,1,1 then you must CLOSE 1,1,1.

Note: don't open a file more than once without closing it first.

If total control is to be made by the device, eg. listing a program onto the printer, then the CMD command must be used. The numbers are as follows:

CMD	DEVICE NUMBER
Printer	4 or 5
Disk	8
Screen	3
Keyboard	Ø
Cassette	1
RS 232	2
Serial bus	4–255

When the device has finished being used it must be CLOSEd by the following method: PRINT # device no: CLOSE device no.

eg.
OPEN 4,4:CMD 4:PRINT"HELLO";:PRINT #4:CLOSE 4

This prints 'HELLO' on the printer and properly closes the channel to the printer and sends control to the screen.

IF a file has been OPENed then it can have information printed on it or data inputed from it.

The printing command is PRINT #. It has the syntax:

PRINT # device no, data.

The data can be in quotes or variables just like the normal PRINT command. When you have finished you must CLOSE the same file that you OPENED.

Information can be PRINT #ed onto any device except the keyboard.

The inputing command is INPUT #.

If a file has been OPENed then information can be INPUT #ed from it.

It has the syntax:

INPUT # device no, data.

The data can be variables or text enclosed in quotes.

When you are finished you must CLOSE the file that you OPENED.

Information can be INPUT #ed from any device except the screen or the printer. Instead of INPUT there is another command as in normal BASIC:

GET

Like in normal BASIC, GET # acts just like GET by getting one bit of data at a time (one variable or letter).

It has the syntax:

GET # device no, variable (string or normal).

As with PRINT # and INPUT #, GET # must also have the channel CLOSED when finished with.

There is a statement which returns the device number of the last device to be used. It is STATUS.

eg.
PRINT STATUS
ex. result. 16....read error.

In a program where there are lots of possible subroutines which can be used, depending on the result of a calculation, for example.

Instead of lots of IF THEN statements we can use an ON GOTO or ON GOSUB command. If the value of the statement is one then the program branches to the first line number in the list, if two then it branches to the second. If the value is less than one or greater than the number of line numbers in the list then the computer goes to the next line of the program (or the next command after the colon).

eg.

```
10 A=1
20 ON A GOSUB 100,110,120,130
30 A=A+1
40 GOTO20
100 PRINT"A=";A
105 RETURN
110 PRINT"A=";A
115 RETURN
```

```
120 PRINT"A=";A
125 RETURN
130 PRINT"LAST IN LIST";A:END
```

Numbers greater than one can be used by using equations as in the following example.

```
10 A=.5:A=A*2
20 ON A GOSUB 100,110,120,130
30 A=A*2
40 GOTO20
100 PRINT"A=";A
105 RETURN
110 PRINT"A=";A
115 RETURN
120 PRINT"A=";A
125 RETURN
130 PRINT"LAST IN LIST";A:END
```

MEMORY CHANGING

These are the most powerful set of commands that the Commodore 64 (or any other computer) possess. They allow machine code to be entered and executed, they can control sound, etc, and they can totally change the computer's behaviour.

The first of these commands is POKE. It has the syntax:

POKE address, number (between 0 and 255)

There are some addresses which cannot be POKEd. These are called ROM. The BASIC interpreter and the character set and the operating system are stored there. All other locations may be POKEd.

POKEing a memory address puts a number into that address and in certain addresses control colour, sound, sprites, high resolution, etc. An address is a location in memory which is unique to any other. They cannot ever change. The values contained in some of them, however, can. Think of a memory address as a letter box that is only big enough for one letter. But that letter can be between zero and 255 units in size. To put another letter into that letterbox you must remove the old one. This is what POKEing does. It acts like the postman. Only it takes a letter away for every new one it delivers.

The addresses that can be POKEd are between zero and 65535. However, addresses between 40960 and 49151 and 57344 and 65535 cannot be POKEd. The character ROM is also unPOKEable and it is between locations 53248 and 57344. (This is paged so these locations can be POKEd. See the section on programmable characters). POKE can be used with variables as well as numbers.

eg.

```
POKE 2040,13 :
POKE A , B
```

WARNING

Make sure the variable A in the example above is always set as POKEing into location zero can cause a crash (never ending loops that you can not stop). The only way to recover from one is to turn the power off and on again... losing your program.

POKE has a complimentary command. It is PEEK. PEEK returns the value contained in a memory address or location. All addresses from zero to 65535 can be PEEKed. PEEK does not alter the contents of the address. It can also be used with numbers or variables.

eg.

```
A = PEEK(65535):PRINT A
```

result 255

```
PRINT PEEK(0)
```

result 47

IF PEEK (1060) = 21 THEN 100. If location 1060 contains 21 then goto 100

There is a command that waits for a location to contain a specific value. It is WAIT. It has two syntaxes:

1. WAIT address, variable
2. WAIT address, variable 1, variable 2

It is not a command that should be used unless you are an experienced programmer, as you could get into a loop that you cannot get out of. It is mainly used in input/output operations. However, one good use for WAIT is for waiting for the user to press a key. To do this more effectively make a line of your program:

POKE 198,0 : WAIT 198 , 1

The program will wait until you press a key and then carry on with the next line.

The way that the WAIT command works is that it takes the value in the memory address and ANDs it with variable one. Then it exclusive ORs (see the machine code section) the number left with variable two. This sounds and is complex but just think of it as a filter which when clear waits until the clear value changes.

Machine code is entered into memory by means of the POKE command. However, we need a method of starting the machine code. There are two ways:

SYS(x) This command 'runs' the machine code from address x.

USR(x) This command runs a machine code subroutine from the addresses stored in addresses 784 and 785 and passes the value of x to the machine code program. SYS is far more powerful and is easier to use. For a fuller explanation see the machine code section of the book.

SECTION 1.2

COLOUR

The Commodore 64 has 16 colours at its disposal. They can be used for the border, the screen, characters, high resolution graphics and sprites. The normal colour resolution (smallest individually colourable spot) on the screen is limited to 40 * 25 (the same as the characters). This section shows only this low resolution as the high resolution will be covered in the section: sprites, high resolution and programmable characters as they allow more colours per unit.

First, we deal with the border and screen colours:

The screen and border colours are controlled separately by the POKE command. They are both in the range 0 to 15. See the table below.

The border is controlled by the location 53280. When the computer is switched on the normal value in this location is 254.

eg.
To change the border to black: POKE 53280,0

The screen is controlled by location 53281. When the computer is switched on or reset, the normal value in this location is 246.

eg.
To change the screen to red: POKE 53281,2

These registers can only show 16 colours but can be POKEd

and PEEKed up to 255. To make sure that the right colour is POKEd or PEEKed it is advisable to put AND 15 after the statement. This filters out excess numbers and gives the true colour.

eg.
POKE 53281,2 AND 15 for a red screen
PRINT PEEK (53281) AND 15 to PEEK the screen.

Characters can be one of 16 colours (as shown in the table). There are three ways of accessing these colours.

1. They are available directly from the keyboard by using the CTRL key (C=) key and the number keys 1 to 8. Using the CTRL key the colours on the front of the number keys can be accessed. Using the Commodore key and the numbers 1 to 8 the second set of colours are accessed. These are shown in the table below. These colour commands can be used in programs by putting them in PRINT statements and they will change the colour of the text following them. They appear in a listing as special symbols.

```
5 POKE53281,0:POKE53280,2:REM BLACK SCRE
EN AND RED BORDER
10 PRINT"EHELLO":REM WHITE
20 PRINT"£YOU↑":REM RED THEN GREEN
25 FORREST=0TO200:NEXTREST
30 GOTO10
```

2. The CHR$(x) statement also changes colours. It is used in conjunction with the PRINT command.

eg.
PRINT CHR$(5);"HELLO";:PRINT CHR$(144)"YOU"

The colours along with their values are in the table.

3. There is a register (location) that contains the current cursor colour (colour printing in at the moment) and it can be

COLOUR CODES TABLE

COLOUR	SCREEN	BORDER	LOOKS LIKE	CHR$(	POKE646,	CTRL &	C=KEY&
BLACK	0	0		144	0	1	---
WHITE	1	1		5	1	2	---
RED	2	2		28	2	3	---
CYAN	3	3		159	3	4	---
PURPLE	4	4		156	4	5	---
GREEN	5	5		30	5	6	---
BLUE	6	6		31	6	7	---
YELLOW	7	7		158	7	8	---
ORANGE	8	8		129	8	---	1
BROWN	9	9		149	9	---	2
LT RED	10	10		150	10	---	3
GREY 1	11	11		151	11	---	4
GREY 2	12	12		152	12	---	5
LT GREEN	13	13		153	13	---	6
LT BLUE	14	14		154	14	---	7
GREY 3	15	15		155	15	---	8

POKEd to change colours. It is location 646. It can be POKEd with the values Ø to 15 as shown in the table (16 – 255 will work but only repeat Ø to 15). This is the most efficient method for changing colour. It is also the fastest.

eg. To print out bars of every colour.

```
10 FORA=0TO15
20 POKE646,A
30 PRINT"R
40 NEXTA
```

If you are having trouble getting good colours on your TV use this program to tune it with. The colours should all be crisp and clear.

SECTION 1.3

ANIMATION

Animation is the movement of characters on the screen, principally for use in games.

As the screen on the Commodore 64 has 40 characters across by 25 characters down, there are 1000 screen locations. The screen is permanently memory mapped (stored in memory) and takes up 1,000 bytes of memory. It is stored from locations 1024, to 2023 inclusive. See the table at end of the section.

There are two ways of producing animation in BASIC. One is using TAB, SPC and the cursor controls but this method is slow and pretty useless so I will not waste time on it. The second, far better method is POKEing to the screen. To put a character onto the screen you POKE the location you want it to appear in with its code from Appendix 2. For example, POKE 1024,1: puts an A in the top left of the screen. Clear the screen and try it. There it is . . . or is it??? Yes, it is there, but you can't see it as it has no colour.

There is a similar part of memory to the screen which is set aside for the colour on the screen. You POKE each location (or the ones you want) with a colour code from 0 to 15. But remember you can only see characters which are not the same as the screen colour. This 'screen' is stored from locations 55296 to 56295 inclusive. See the table at end of the section.

To colour the A you POKEd into 1024, type POKE 55296,1 (or any colour you want). The A should now be displayed in white.

The colour map is laid out exactly the same as the character map (screen) except that it is 54272 locations further on. The easiest method to colour a character that you have POKEd onto the screen is to do the following:

POKE location , character : POKE location + 54272 , colour

To find out what character is on the screen or what colour it is we use the PEEK( ) statement.

eg.
PRINT PEEK (1Ø24)

If the A was still in 1Ø24 then 1 would be PRINTed, otherwise 32 would be returned (32 is the code for a space). If we PEEKed the colour it would give 1.

eg.
PRINT PEEK (55296) or PRINT PEEK (1024 + 54272)

The answer should be 1. The following short game demonstrates the use of POKEing and PEEKing the screen.

Use the M key to move right and the Z key to move left. You have to dodge the asteroids (full stops) and ram the oncoming ships (A). Good luck.

```
5 PRINT"s"
10 POKE53281,0:POKE53280,0:FORA=0TO24:PR
INT:NEXT:B=1504:C=20
20 A=RND(1)*40:POKE1984+A,46:POKE1984+A+
54272,RND((1)*14)+1
25 IFRND(1)>.7THENP=RND(1)*40:POKEP+1984
,1:POKEP+1984+54272,7
30 GETA$:IFA$="Z"ANDC>0THENPOKEB+C,32:C=
C-1
35 IFA$="M"ANDC<40THENPOKEB+C,32:C=C+1
```

```
40 POKEB+C,32
50 PRINT
60 IFPEEK(B+C)=46THEN120
70 IFPEEK(B+C)=1THENGOSUB200
90 POKEB+C,22:POKEB+C+54272,9
100 GOTO20
120 PRINT"s":POKE53281,2:PRINT"YOU HIT A
N ASTEROID"
125 PRINT"YOU SCORED ";S;"POINTS":S=0
130 PRINT"DO YOU WANT TO TRY AGAIN?"
140 GETA$:IFA$="Y"ORA$="N"THEN150
145 GOTO140
150 IFA$="N"THENEND
160 IFA$="Y"THENRUN
200 FORA=0TO16:POKE53280,A:NEXT:S=S+10:R
ETURN
```

As you can see from the program (lines 30–50) before you move you must erase your last position. This is achieved by POKEing a 32 (space) onto your position and *then* moving.

eg.
POKE location, 32 :move character: POKE location , character.

If you feel easier by positioning characters on the screen in columns and rows, you can still POKE to the screen using co-ordinates using the following formula:

POKE 1024 + column + (40 * row) , character
(Columns are from zero to 39 and rows from 0 to 24.)

Use the same 'X and Y co-ordinate method for the colour, replacing 1024 with 55296. Then to move, update row and column variables respectively.

Below are the screen and colour maps for the Commodore 64.

SCREEN MEMORY MAP

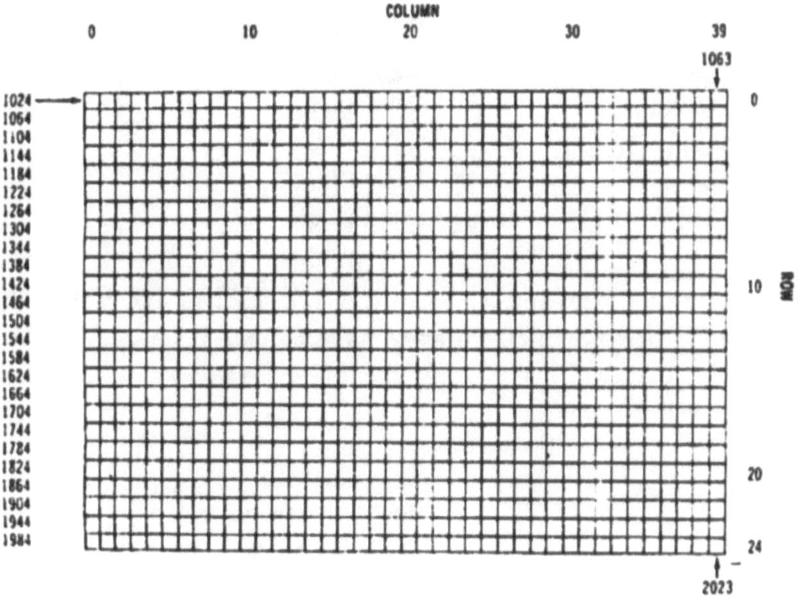

COLOUR MEMORY MAP

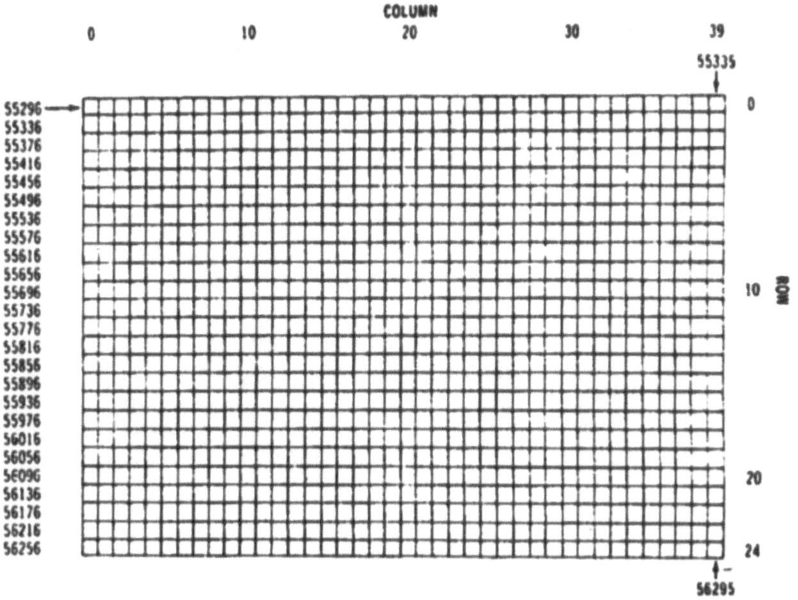

The values to POKE into the above colour locations are as follows:

Ø	BLACK	1	WHITE
2	RED	3	CYAN
4	PURPLE	5	GREEN
6	BLUE	7	YELLOW
8	ORANGE	9	BROWN
1Ø	LIGHT RED	11	GREY 1
12	GREY 2	13	LIGHT GREEN
14	LIGHT BLUE	15	GREY 3

For example, to make the colour of location 1Ø24 (left-hand top corner of the screen) yellow type POKE 55296,7.

SECTION 1.4

MUSIC AND SOUND SYNTHESIS

The sound output from the Commodore 64 is the best from any computer I have ever heard. You can totally control the individual 'voices' to make them sound like any sound, note or voice that you want. The Commodore 64 has three individual voices and one, two or all three of them can be played simultaneously. The sound can even be fed to an external amplifier to give a really professional sound.

In this section I will refer only to voice 1 unless stated. However, the principles for the other two voices are exactly the same. Simply substitute the numbers for the voice you want to use in place of the ones I use for voice 1. These numbers are in the table at the end of this chapter and in Appendix 8.

Sound on the Commodore 64 is controlled by the POKE command. It changes values in the SID (Sound Interface Device) chip which is the sound microprocessor in the 64. It has 27 different registers which control the three voices. Before you create any sound you need to set various parameters. They should be done in the following order.

1: VOLUME. This is a master control which controls the volume of all three voices at once. It is in the range Ø (no volume — off) to 15 (full volume) values higher than 15 do other things and these will be explained later. The register to POKE is 54296. If you want to set volume to full, type POKE 54296,15.

2: ATTACK/DECAY — This is set separately for all three voices. It is the setting which controls the rate at which the note 'hits you' and rises to its peak volume. Both values are controlled by one register per voice. The attack is controlled by the left-hand four bits (eight bits = one byte) and the decay is controlled by the right-hand four. The register to POKE is 54277.

ATTACK/	POKE	ATK4	ATK3	ATK2	ATK1	DEC4	DEC3	DEC2	- DEC1
DECAY	54277	128	64	32	16	8	4	2	1

The larger the number the longer the time till the note reaches its peak. To get longer attack or decays or both add the respective bits together.

To get a long attack but no decay add bits 8,7,6 and 5 together.

eg.
128 + 64 + 32 + 16 = 240
So POKE 54277 , 240.

Or to get a medium attack and a medium decay, try:

64 + 32 + 4 + 2
POKE 54277, 102.

3: SUSTAIN/RELEASE — This is set separately for each voice. It is the setting which controls how long a note stays at its maximum and how fast it decays to zero. Both settings are controlled by one register per voice. The sustain is controlled by the left-hand four bits and the release by the right-hand four.

The register to POKE is 54278.

SUSTAIN/	SUS4	SUS3	SUS2	SUS1	REL4	REL3	REL2	REL1	POKE
RELEASE	128	64	32	16	8	4	2	1	54278

The longer the sustain the longer the note takes to fade to zero. To get long sustains or releases or both, add the bits in the above table and POKE into 54278.

eg.
To program a note that will play indefinitely, use no release:
POKE 54278,128

Or a note that releases as soon as it reaches its peak:
POKE 54278,7

4: WAVEFORM – Each voice on the Commodore 64 can have one of four possible waveforms: triangle, sawtooth, pulse, or noise.

You POKE register 54276 with the value for the waveform you want.

WAVEFORM	POKE	TRIANGLE	SAWTOOTH	PULSE	NOISE
	54278	17	33	65	129

Changing the waveform can drastically change the sound produced.

For example, to create a sawtooth waveform for voice 1, type

POKE 54276,33.

5: HIGH FREQUENCY/LOW FREQUENCY – Each note you play on the Commodore 64 requires two values to be POKEd. The low frequency is nearly always the higher number. The *low frequency register is 54272. The high frequency* register is 54273.

For example, to play a C (if all the parameters have been set), type

POKE 54272,75 : POKE 54273,34

These two registers are used to play music or make noises on the triangle, sawtooth or noise waveforms. The pulse waveform has two different registers to be POKEd.

5/b: (Pulse waveform only) — These are the low and high frequency registers for the pulse waveform. High frequency pulse is register 54275 which is POKEd with a number from 0 to 15. Low frequency pulse is register 54274 which is POKEd with a number from 0 to 255.

For example, to play a note first POKE 54276,65 (pulse set) then type

POKE 54274,150:POKE 54275,7

To play a scale on voice 1, use the following program. It uses the triangle, sawtooth and noise waveforms.

```
5 WF=17
10 POKE54296,15:POKE54277,8:POKE54278,35
:REM VOL AND ADSR
15 FORX=0TO2
20 POKE54276,WF:REM WAVEFORM
22 IFWF=17THENWF=33:GOTO30:REM CHANGE WA
VEFORMS
24 IFWF=33THENWF=129:REM CHANGE WAVEFORM
S
30 FORA=0TO8:READL,H:REM NOTE DATA
40 POKE54272,L:POKE54273,H:REM MAKE NOTE
50 FORR=1TO300:NEXTR:REM PAUSE
60 NEXTA:REM GO BACK & DO NEXT NOTE
65 RESTORE:NEXTX
70 DATA75,34,126,38,52,43,198,45,97,51,1
72,57,188,64,149,68,0,0
```

There are also filters available for each voice. There is a high pass filter, a low pass filter and a band pass filter. They are all controlled by location 54296. The low pass filter is turned on by POKEing 54296 with 16. The high pass filter is turned on by

POKEing 54296 with 64. The band pass filter is turned on by POKEing 54296 with 32. (To turn on volumes as well add 15 to values). There are also filters which 'turn off' each voice; however, when you switch these filters off the voice sounds the same as before you filtered it. This is location 54295.

To filter voice 3 POKE 54295,4
To filter voice 2 POKE 54295,2
To filter voice 1 POKE 54295,1

You can also add resonance to the three voices. To do this you POKE 54295 with the respective bit (four to seven) ie. 8 to 128. You can add these values together to filter, each or all voices. For example to filter both voice 1 and 2 type POKE 54295,3.

SECTION 1.5

PROGRAMMABLE CHARACTERS

Every and any character that can be displayed on the screen can be changed or reprogrammed to represent anything else, say, a space invader or a Greek character set.

Before trying to program characters it is necessary to understand how the normal character set is stored in memory.

A character is made up of 64 dots, eight dots across by eight dots down. Each dot is a bit of computer memory and as eight bits are one byte a character takes up 8 * 1 byte which means that each character needs eight bytes to store.

Here is how the letter 'A' is stored in memory. Each bit can have one of two values 1 or Ø. The bits are then added together to get a decimal number for each of the eight rows. For the A they are as follows:

```
Ø Ø Ø 1 1 Ø Ø Ø         24
Ø Ø 1 1 1 1 Ø Ø         60
Ø 1 1 Ø Ø 1 1 Ø        102
Ø 1 1 1 1 1 1 Ø        126
Ø 1 1 Ø Ø 1 1 Ø        102
Ø 1 1 Ø Ø 1 1 Ø        102
Ø 1 1 Ø Ø 1 1 Ø        102
Ø Ø Ø Ø Ø Ø Ø Ø        Ø
```

The data for an A is therefore 24,6Ø,1Ø2,126,1Ø2,1Ø2,1Ø2,Ø.

Once the data for a character has been calculated it needs to be put into memory. There are 65 1K blocks set aside for this purpose. However, not all of these can be used for this purpose. They start at Ø and go up to 65536 but the first sensible one to use is 8192 (the eighth 1K block). This allows 6K of BASIC to be before your characters. If you need more than 6K then must move your characters to a higher block. So, if you need 16K then start your characters at 18432 (decimal).

The video chip can only 'see' 16K at one time. Therefore a way is needed to allow it to 'see' the whole of the 64's memory. This method is called 'Banking'. It is controlled by location 56576.

To change banks type the following in direct mode or in a program...

```
POKE 56576,(PEEK(56576)AND 254)OR A
```

Where A has the value from the following table.

Value of A	Bits	Bank	Starting Location	Video Chip Range
Ø	ØØ	3	49152	($ CØØØ–$ FFFF)
1	Ø1	2	32768	($ 8ØØØ–$ BFFF)
2	1Ø	1	16384	($ 4ØØØ–$ 7FFF)
3	11	Ø	Ø	($ ØØØØ– 3FFFF) (DEFAULT VALUE) (Normal)

The screen memory is the first 1ØØØ bytes in each bank, or is the colour memory in high resolution.

For the A character we put it into memory as follows (the A starts at 82ØØ as the @ character is from 8192 to 8199):

```
10 FORA=0TO7
20 READB:POKE8200+A,B
30 NEXTA
40 DATA 24,60,102,126,102,102,102,0
```

Run this program. Then press the A key. You just get a normal A. In order to see our programmed character we need to tell the computer where to look for its character data. For data starting at 8192 type

POKE 53272,24.

The screen will now go strange. Press the A key. You should see a normal A. But if you press any other key or list the program, you will see lines. This is because only the A has been programmed to look like something that you want. Address 53272 is the screen/character memory pointer. If you want to have your character data higher up memory, then add one to 24 for every 1K that you move your data up memory. Then POKE this number into address 53272. So, if you want to start your data at 16384 (16K) type

POKE 53272,32 (24 + 8)

To program different characters you need to utilise the screen character codes in Appendix 10, ie. C=3. The formula for calculating the starting position for a specific character is as follows. I will use 8192 as my starting position.

Character data address = 8192 + (screen code * 8) to 8192 + (screen code * 8) +7.

To program more than one character, put all the data together and then use one or more FOR NEXT Loops to POKE the data in. The following program produces three characters re-programmed.

```
10 FORA=0TO23
20 READB:POKE8192+A,B:NEXT
30 POKE53272,24
```

```
40 PRINT"]]●]]AB"
50 DATA 129,189,165,255,126,36,66,36,0,2
,15,25,57,255,24,24,0,64,240,152,164,255
60 DATA12,12
```

Sometimes you may want most, some or all of the standard characters in your new character set. Instead of programming each one by hand there is a quick and efficient method which can program some, most or all of the character set into RAM. The following program copies all the character set into RAM. If you want less than the whole set in RAM then adjust the FOR NEXT loop in line 30.

```
10 POKE56334,PEEK(56334)AND254
20 POKE1,PEEK(1)AND251
30 FORA=0TO2048:POKE8192+A,PEEK(53248+A)
:NEXT
40 POKE1,PEEK(1)OR4
50 POKE56334,PEEK(56334)OR1
60 POKE53272,24
```

There are a few important things in this program which are hard to understand due to the architecture of the Commodore 64. Here is a line by line breakdown of the program.

Line 10: This command disables the interrupt. This allows the character ROM to be used.
Line 20: This pages in the character ROM on top of the video chip.
Line 30: This copies the character ROM to RAM.
Line 40: This resets the interrupt.
Line 50: This pages the video chip back in and the character ROM out.
Line 60: This puts the 64 into 'programmed character mode'.

DOUBLE CHARACTERS

Unluckily there is no simple way to create and use double height characters on the 64, but it is still possible. The method involves POKEing the data into two consecutive bytes.

For example, a normal height A looks like this:

```
Ø Ø Ø 1 1 Ø Ø Ø
Ø Ø 1 1 1 1 Ø Ø
Ø 1 1 Ø Ø 1 1 Ø
Ø 1 1 1 1 1 1 Ø
Ø 1 1 Ø Ø 1 1 Ø
Ø 1 1 Ø Ø 1 1 Ø
Ø 1 1 Ø Ø 1 1 Ø
Ø Ø Ø Ø Ø Ø Ø Ø
```

And a double height A looks like this:

```
Ø Ø Ø 1 1 Ø Ø Ø
Ø Ø Ø 1 1 Ø Ø Ø
Ø Ø 1 1 1 1 Ø Ø
Ø Ø 1 1 1 1 Ø Ø
Ø 1 1 Ø Ø 1 1 Ø
Ø 1 1 Ø Ø 1 1 Ø
Ø 1 1 1 1 1 1 Ø
Ø 1 1 1 1 1 1 Ø
Ø 1 1 Ø Ø 1 1 Ø
Ø 1 1 Ø Ø 1 1 Ø
Ø 1 1 Ø Ø 1 1 Ø
Ø 1 1 Ø Ø 1 Ø Ø
Ø 1 1 Ø Ø 1 1 Ø
Ø 1 1 Ø Ø 1 1 Ø
Ø Ø Ø Ø Ø Ø Ø Ø
Ø Ø Ø Ø Ø Ø Ø Ø
```

In other words each byte of the normal height character is repeated.

In the following program a double height A is programmed.

```
10 FORA=0TO15STEP2:READB:POKE8192+A,B:PO
KE8193+A,B:NEXT
20 PRINT"s@":PRINT"A"
30 POKE53272,24
40 DATA24,60,102,126,102,102,102,0
```

You will now see a double height A in the top left-hand corner of the screen. One double height character takes two normal height characters to display. Because of this it is really only useful to use double height characters in something like the title of a program.

The following program programmes the entire character set (up to code 128) to double height characters. The characters are made up in the same way as the previous program except that instead of data the character ROM is being used.

```
10 POKE56334,PEEK(56334)AND254
20 POKE1,PEEK(1)AND251
30 FORA=0TO512STEP2:POKE8192+A,PEEK(5324
8+A/2):POKE8193+A,PEEK(53248+A/2)
40 NEXT
50 POKE56334,PEEK(56334)OR1
60 POKE1,PEEK(1)OR4
70 POKE53272,24
80 PRINT"s@1B1D1F1H1J1L1N1P1R1T1V1X1Z"
90 PRINT"A1C1E1G1I1K1M1O1Q1S1U1W1Y1["
```

MULTICOLOUR CHARACTERS

So far we have only programmed characters in one colour. However, they can be multicolour having four colours out of 16 in each eight by eight character.

The procedure for creating these characters is basically the same as characters in one colour but there are some differences.

The horizontal resolution of the characters is half that of one colour characters but the extra colour makes this sacrifice worthwhile. The colour resolution of the characters is still as high as on the arcade machines. The dot is double the width because the method used to programme the dots in more than one colour involves bit pairs. Instead of an 'on' dot being a one and an 'off' dot being zero there are four different combinations for on/off. These are 01, 10, 11, and 00. So, instead of putting a one where you want an 'on' dot you put an 11 or an 01 or a 10 depending on which colour you want the dot to be. For an 'off' dot you put a 00.

These 'bit pairs' (pair of binary bits) stand for:

BIT PAIR	MEANING
00	Screen colour
11	Character colour
01	Multicolour 1
10	Multicolour 2

To design your multicoloured character you need a four by eight grid. But the four across are all two bits in size.

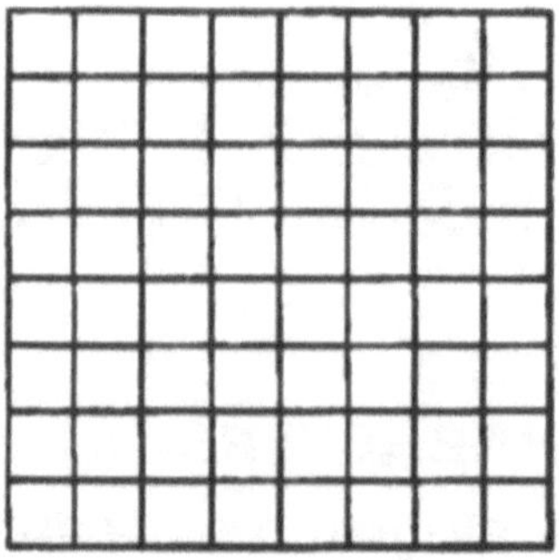

bits 76543210

In multicoloured character mode a man could look like:

The binary data for this character is:

00111100, 00011000, 11011011, 00011000, 00111100, 00100100, 00100100, 11000011.

And the decimal data is:

60,24,219,24,60,36,36,195

You just POKE the data into memory as if the characters were normal characters using the same locations (8192, for example). Once you have done this you enter the programmable character mode with:

POKE 53272,24 (or any number you have already specified by the position of your character data).

Then you have put the 64 into a multicolour mode. This is done with:

POKE 53270,152

If your cursor colour is betwen black and yellow (codes 0 to 7) then you will notice no difference. Press the C= key and 1. Now type something. See the characters are now in four colours. You will notice that the characters are almost unreadable. This is because multicolour characters depend on the various 'bit pairs', and normal characters are programmed to look good in one colour. Therefore, multicolour is really only useful for user-defined (or programmed) characters. We are not limited to the colours you see now. All of the three character colours are individually programmable. Each location controls a different 'bit pair' combination.

The character colour or the colour screen POKE are controlled by printing in a colour code that is greater than seven (orange) by PRINT CHR$ (or POKEing location 646 or POKEing the screen colour POKE). This colours the dots with bit pairs of 11.

Multicolour 1 is controlled by POKEing location 53282. This is

POKEd with a number between Ø and 255. This location control the colour of 'bit pairs'.

CODE Ø1

Multicolour 2 is controlled by location 53283. It is POKEd with numbers between Ø and 255. This location controls 'bit pairs' with code 1Ø. The fourth colour is the background or screen colour and can only be altered by changing the screen colour with location 53281.

This is a program which programs a ghost in four colours and moves it about the screen.

```
5 PRINT"s"
10 POKE53270,152:POKE53282,7:POKE53283,2
:POKE53281,0:POKE53280,0:POKE53272,24
20 FORA=0TO31:READB:POKE8192+A,B:NEXT:FO
RA=0TO7:POKE8192+32*8+A,0:NEXT
30 FORA=0TO39:POKE55296+A,13:POKE55297+A
,13:POKE55336+A,13:POKE55337+A,13
40 POKE1024+A,0:POKE1025+A,1:POKE1064+A,
2:POKE1065+A,3
50 POKE1023+A,32:POKE1063+A,32
60 NEXT
70 POKE1103,32:POKE1104,32:GOTO30
2000 DATA0,3,15,63,255,253,253,253,0,192
,240,252,255,117,182,117,255,255,255,255
2010 DATA255,255,51,51,255,255,255,255,2
55,255,51,51
```

It can therefore be seen that when you are designing your characters you need to use 'bit pairs' to get the desired effect.

EXTENDED COLOUR MODE

This mode allows you to colour the background as well as the foreground of a character that you can display on the screen. it is turned on with the command,

POKE 53265,91.

For the first 64 characters the background colour (which can be from Ø to 16) the screen forms the background. For the second 64 (65–127) location 53282 controls the colour. For the third 64 (128–191) location 53283 controls the colour. For the fourth 64 (192–255) location 53284 controls the colour. As can be seen, only characters with a code greater than 64 can have their background individually controlled. Thus, it is really only useful in a program. Warning — do not use this mode in conjunction with the multicolour character mode as it does not work. To turn off the extended colour mode, type

POKE 53265,27.

The following short program demonstrates the use of such characters:

```
10 poke53265,91:poke53281,0:poke53282,5:
poke53283,2:poke53284,8:print"S"
20 print"sqqq^^^^^not extended colour mo
de"
30 print"qqq^^^^rextended colour mode":r
em just type letters
40 print"qqq^^^^ENTENDED COLOUR MODE":re
m type letters with the shift key presse
d"
50 print"qqq^^^^rENTENDED COLOUR MODE":r
em type letters with the shift key press
ed"
```

SECTION 1.6

SPRITES

This is probably the most useful and effective graphic capability on the Commodore 64. The sprites are far better than any on any other computer that have this feature. Their quality is as good as, or better than, arcade video games.

A sprite is an object that is 24 dots across by 21 down. It can be coloured in any four of 16 colours at once. It is moveable one pixel (dot) at a time and automatically detects collisions with other sprites or screen characters and informs the video processor that it has done so.

Sprites are generated by and controlled by the POKE command. A sprite is created on a 24 by 21 dot grid (three bytes across by 21 down). You add up the 'on' bits just as you would in programmable characters, except that there are three bytes across (instead of one) and 21 down (instead of eight) ie. 21 groups of three bytes.

You add the bits together for each byte, going across then down. So you do row 1s three bytes then row 2s, etc.

I will create a sprite in one colour to start with.

To turn this picture into a strawberry (or any sprite) you add up the bits and form a list of 63 bits of data.

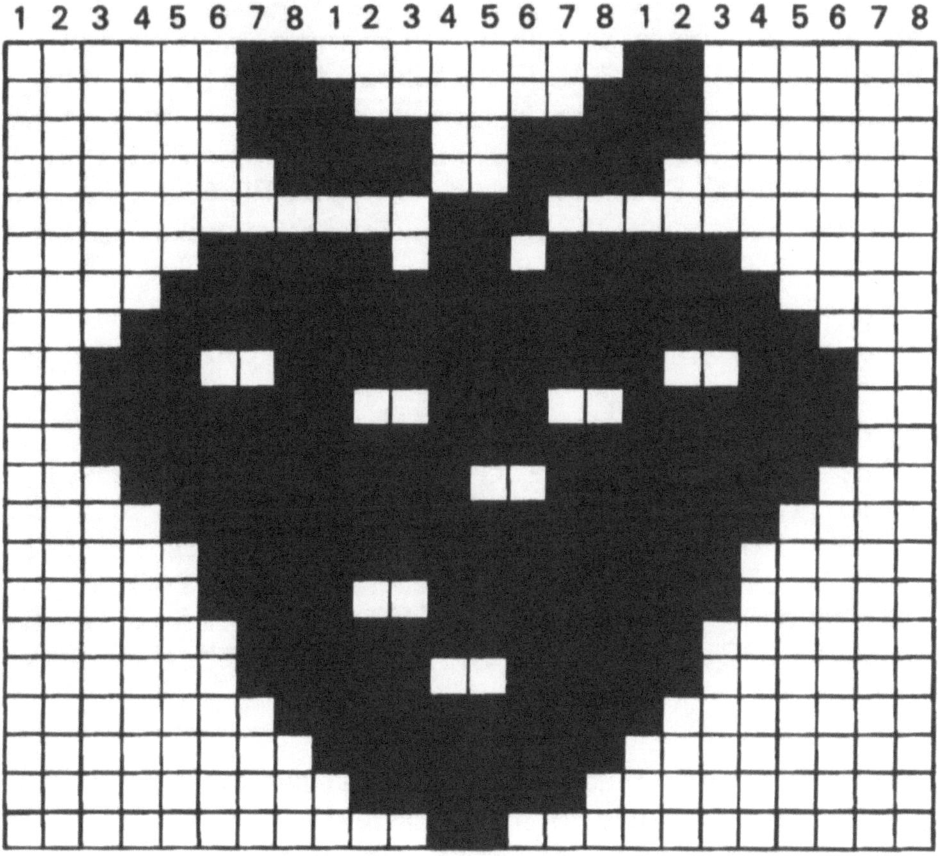

So the data is as follows:

```
1000 DATA 3,0,192,3,129,192,3,231,128,1,
231,128,0,24,0,7,219,0,15,219,224
1010 DATA 15,255,240,31,255,240,57,255,1
56,63,158,252,63,255,252,21,243,248,15
1020 DATA 255,240,15,159,208,3,255,192,3
,221,192,1,255,128,0,255,0,0,126,0,0,24,
0
```

Right. Now sprites have to be stored in memory like programmable characters. Each one is stored in a block of memory that is 64 bytes long. These blocks start at location 0 and go up to 16320. However, the first block that can safely be used is 13, ie address 832 onwards. Addresses 1024 to 2047 are also unuseable as the screen is there. So you can use locations 2048 to 16320. To calculate what the blocks starting address is, use this simple formula: address = block * 64. Blocks up to 255 can be used. For example, block 250 starts at location 16000. So using the data above we POKE the data into a block. I will use 13 (832 –) so our program looks like this.

```
10 FORA=0TO62
20 READB:POKE832+A,B:REM PUT SPRITE DATA
 INTO MEMORY
30 NEXT
1000 DATA 3,0,192,3,129,192,3,231,128,1,
231,128,0,24,0,7,219,0,15,219,224
1010 DATA 15,255,240,31,255,240,57,255,1
56,63,156,252,63,255,252,21,243,248,15
1020 DATA 255,240,15,159,208,3,255,192,3
,221,192,1,255,128,0,255,0,0,126,0,0,24,
0
```

Now our sprite is created but we can't see it. There are two reasons for this — it is not turned on and it is off the screen. For this we need to go to the video chip which starts at location 53248.

To make the numbers easier to remember I will assign the beginning of the video chip to a variable, ie V = 53248.

I will refer to registers in the video chip as the variable plus a number, ie V + 4.

NOTE

When POKEing the variables make sure that V is set at 53248. Otherwise the examples will not work and the computer might crash (go into an endless loop).

To turn on a sprite we use register V + 21. Each sprite is turned on by its respective bit:

Sprite Ø = bit 1
Sprite 1 = bit 2
and so on to:
Sprite 7 = bit 8

Each bit has its own specific value.

sprite number	7	6	5	4	3	2	1	Ø
bit number	8	7	6	5	4	3	2	1
value	128	64	32	16	8	4	2	1

For example, Sprite Ø is turned on with POKE V+21,1 (or POKE 53269,1). To turn on more than one sprite at a time you add their 'bit values' together.

eg.
To turn on sprites Ø,1,2 and 7, add 1+2+4+128=135 so type POKE V+21,135.

Now we have to tell the video chip where to find the data for our sprite. This is handled by eight locations which are just before BASIC RAM (where BASIC or machine code programs are written). You POKE the respective location with the block number in which the sprite data is stored. The addresses are as follows:

Sprite Ø 2Ø4Ø
Sprite 1 .... 2Ø41
Sprite 2 .... 2Ø42
Sprite 3 .... 2Ø43
Sprite 4 .... 2Ø44
Sprite 5 .... 2Ø45
Sprite 6 .... 2Ø46
Sprite 7 .... 2Ø47

eg.

To tell the video chip that the data for sprite 2 is in block (13(832 –) type:

```
POKE 2042,13
```

If the data for sprites 3 and 5 are in block 13 then type:

```
POKE 2043,13:POKE 2045,13
```

Now turn on sprite 2 with POKE V+21,4. To tell the video chip that the data for our strawberry is in block 13, type POKE 2042,13 (I am making the sprite 2 the strawberry).

However, we still cannot see our sprite. We need to position the sprite onto the sceen. Each sprite has an X co-ordinate and a Y co-ordinate. These are controlled by two registers for each sprite. For sprite 2, the X co-ordinate register is V + 4, (or 53252). It can be POKEd from 0 to 255. Try POKE V+4,100.

We still cannot see our sprite. This is because the sprite is still, above the screen. The register for the Y co-ordinate is V+5. It can be POKEd with 0 to 255. Try POKE V + 5,100. There it is.

If you can't see the sprite then it is probably the same colour as the screen. Try changing the screen colour or the sprite colour

The registers to change the sprite's colour are V+39 to V+46 for sprites 0 to 7 respectively. To change the colour of a sprite, POKE the respective register with a number from 0 to 15. (The colour codes for sprites are the same as the screen or character codes.) So, to change the colour of the strawberry to red type POKE V+41,2 (or 10 if you want a light red strawberry.)

Using the horizontal(X) and/or vertical(Y) registers, we can move the sprite anywhere on the screen. I know a strawberry does not fly but anyway here goes.

```
10 FORA=0TO62
20 READB:POKE832+A,B:REM PUT SPRITE DATA
 INTO MEMORY
30 NEXT
40 POKE2042,13:PRINT"s":V=53248:POKEV+21
,4:POKEV+41,2:POKE53281,0:POKE53280,0
50 POKEV+5,200:FORX=0TO255:POKEV+4,X:NEX
T
60 FORY=200TO0STEP-1:POKEV+5,Y:NEXT
70 POKEV+4,0:GOTO50
1000 DATA 3,0,192,3,129,192,3,231,128,1,
231,128,0,24,0,7,219,0,15,219,224
1010 DATA 15,255,240,31,255,240,57,255,1
56,63,156,252,63,255,252,21,243,248,15
1020 DATA 255,240,15,159,208,3,255,192,3
,221,192,1,255,128,0,255,0,0,126,0,0,24,
0
```

In this program it is lines 50 and 60 that do the work. They use a loop to POKE the co-ordinates into the X and Y registers. *But wait*!!!!! Why is the sprite not going all the way across the screen???

This is because the Commodore 64's screen is 320 dots across by 200 dots down and a register can only hold a number between 0 and 255. To get around this problem the 64 has a special register which allows us to use the rest of the screen. It moves the sprite to the 256th dot across. Then you use the X co-ordinate register (V + 4) to move the sprite. The register is V + 16. You POKE this register with the bit value for the sprite that you want to go to the end of the screen. For example, POKE V + 16,129 puts sprites 0 and 7 to the 256th dot. When your sprite is finished in this part of the screen you *must* put it back by resetting V + 16 to zero (or the value contained in V + 16 minus the bit value for the sprite you are removing). For example, if if contained 129 and you remove only sprite 7 then you would POKE V + 16,1. To make the last program make the sprite move to the end of the screen change line 50 to:

```
50 POKEV+5,200:FORX=0TO255:POKEV+4,X:NEX
T:POKEV+16,4:FORX=0TO63:POKEV+4,X:NEXT
```

And line 60 to:

```
60 FORY=200TO0STEP-1:POKEV+5,Y:NEXT:POKE
V+16,0
```

A sprite does not have to remain 24 dots across by 21 dots down. It can be doubled in width, height or both, by two simple commands.

The sprite is created normally and can then be expanded at will. The resolution of dot is halved (ie the dot becomes double the dimensions).

The register for horizontal expansion is V + 29(53277). You POKE V + 29 with the bit value for the sprite/s that you want to expand. To return the sprite to normal, you POKE V + 23 with zero or the value in the register minus the sprites bit value.

Stop the sprite program with the strawberry on the screen. Type POKE V + 29,4 and you should see that your sprite has now doubled in width.

The register for vertical expansion is V + 23(53271). You POKE this register with the sprite's bit value just as you did for horizontal expansion. To return your sprite to normal you type POKE V + 23,0 (or the total in the register minus the sprites bit value). Try POKE V + 23,4 and you should find that the sprite has now doubled in height.

The vertical and horizontal registers can be used separately or together. So you can have normal, double width, double height and both double height and width sprites on the screen at once.

Now back to the old program. But *do not* turn the 64 off. Type in the following program... and run it.

```
1 PRINT"s":POKE53281,0:POKE53280,0
10 POKE2040,13:POKE2041,13:POKE2042,13:P
OKE2043,13
20 V=53248:POKEV+21,255:POKEV+39,2:POKEV
+40,3:POKEV+41,10:POKEV+42,12
30 POKEV,100:POKEV+2,200:POKEV+4,100:POK
EV+6,200
40 POKEV+1,100:POKEV+3,100:POKEV+5,100:P
OKEV+7,200
50 POKEV+23,6:POKEV+29,3
60 FORR=0TO200:NEXT:REM REST
70 POKEV+23,1:POKEV+29,4
80 FORR=0TO200:NEXT
90 GOTO50
```

Stop the program. Clear the screen and list the program. Notice that the sprites cover the characters. Sprites can be made to go under text or graphics. The register V+27 controls this. As with other control registers, you POKE it with the respective bit values for each sprite. Type POKE V+27,3. See the sprites go behind the characters. When you want the characters to go back in front of the characters you can, for example, reset this register with POKE V+27,Ø.

A sprite can detect collisions with other sprites or with screen characters. There are two registers which control this.

Firstly sprite/sprite collision. This is register V+3Ø(53278). If a sprite hite another sprite then the two sprites bits are set.

eg.
If sprite Ø hits sprite 2 then register V+3Ø would contain five, ie with bits set.

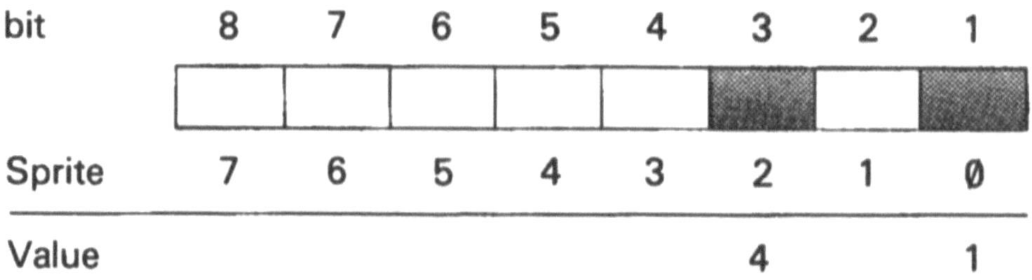

If all eight sprites were in collision then all eight bits would be set and V + 3Ø would contain 255.

New the old program but do not turn the machine off. (This is because you will need the sprite data.) Try this program using two strawberries.

```
10 FORA=0TO24:PRINT"1
          1":NEXT
20 POKE2042,13:POKE2040,13:V=53248
30 POKEV+21,5:POKEV+41,10:POKEV+39,5
40 POKEV+1,100:POKEV+5,100:D=1
50 POKEV,X+30:POKEV+4,255-X:X=X+D
55 IFPEEK(V+31)=4ORPEEK(V+31)=5THEND=1
60 IFPEEK(V+30)=5THEND=-1
70 GOTO50
```

There is a register which detects whether a sprite has hit a background character — it is V + 31. This location has the sprites respective bit set if that sprite hits a character. For example, if both sprites 1 and 2 hit the background then V + 31 would contain six.

Unluckily there is no simple way of telling which character the sprite has hit. The only way is to use a subroutine like the one below.

```
10000 C=INT(1024+(PEEK(SPRITE Y REG))/8+
(PEEK(SPRITE X REG))/8)
10010 IFPEEK(V+16)=SPRITE'S BIT THENC=C+
32
10020 CH=C
10030 REM CHARACTER IS IN CH
```

MULTICOLOUR SPRITES

Muticolour mode allows each sprite to have four colours in its at once.

The principal for programming them is basically the same as multicolour characters in that they have half the horizontal resolution and the use of bit pairs determines their colours.

For multicolour sprites the following bit pairs control these colours:

00 Screen colour
10 Sprite foreground colour
11 multicolour 2
01 multicolour 1

When you have created your sprite and coloured it in multicolour, it retains its colours unless you change them or press RUN/STOP and RESTORE.

To create a multicolour sprite we first need to turn on multicolour mode. This is done by register V + 28. Each sprite is put into multicolour mode by POKEing V + 28 with the sprites respective bit. For example, to put sprites 0 and 2 into multicolour add 1 + 4 = 5.

So POKE V + 28,5.

There are certain registers for controlling the colour of each sprite. The foreground (10 bit pair) is controlled by registers V + 39 to V + 46, for sprites 0 to 7 respectively. For example, POKE V + 41,2 makes the foreground of sprite 2 red.

The register V + 37 controls the colour (multicolour 1) with bit pair Ø1. This register controls the multicolour 1 for all eight sprites. So, if yo type POKE V + 37,2 any multicoloured sprite that is on will have any 0Ø1 bit pairs coloured red.

The register V + 38 controls the colour (multicolour 2) with bit pair 11. This register controls the multicolour for all eight sprites. So, if you type POKE V + 38,5 any multicoloured sprite that is on will have any 11 bit pairs coloured green.

So you design a sprite with bit pairs as follows. I will design a strawberry.

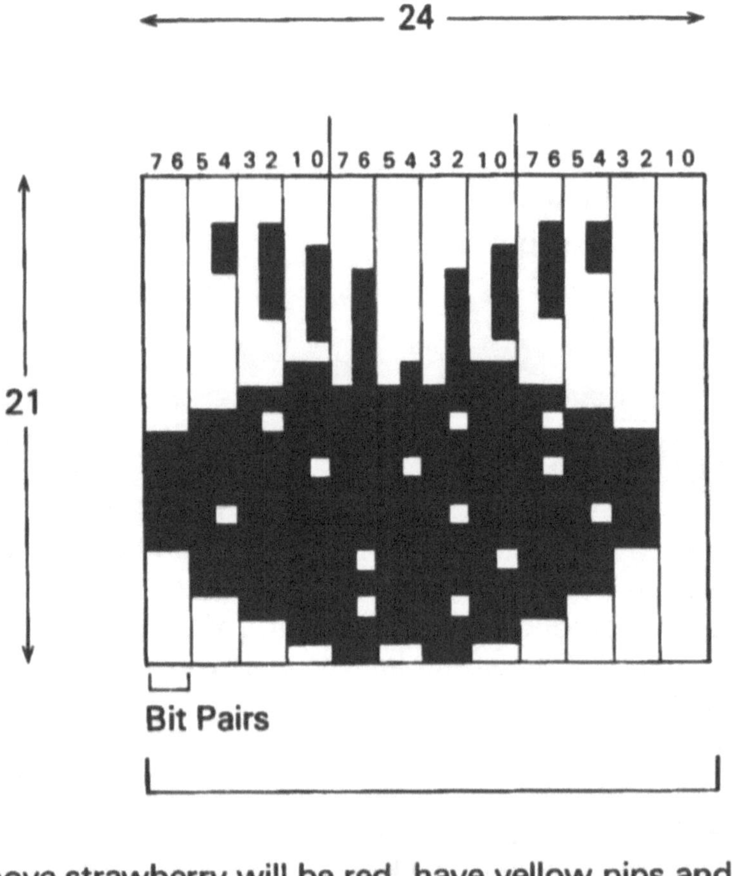

The above strawberry will be red, have yellow pips and green leaves. Therefore, the red will be bit pair 11, the yellow 1Ø and the green Ø1.

Now the data for the above sprite is added up an POKEd into memory.

```
10 FORA=0TO62:READB:POKE832+A,B:NEXT
100 DATA 0,0,0,0,0,0,20,0,80,21,1,80,5,6
9,64,5,69,64,1,69,0,0,68,0,3,87,0,15,255
110 DATA 192,59,251,176,255,255,252,254,
239,188,255,255,252,239,251,236,255,255
120 DATA 252,63,190,240,63,255,240,15,18
7,192,3,255,0,0,204,0
```

Now we need to turn on our sprite (I will use sprite 2):

POKE V+21,4

Now turn on multicolour mode for sprite 2:

POKE V+28,4

Put the sprite onto the screen:

POKE V+4,100:POKE V+5,100

Now we colour the sprite — to colour the sprite strawberry red:

POKE V+38,2

Now we make the leaves green:

POKE V+37,5

Now we make the pips yellow:

POKE V+41,7

Now we have a sprite in four colours (the background being the fourth).

The following program generates all eight sprites in four colours and moves them about the screen.

```
5 PRINT"s"
10 V=53248:POKEV+21,255:POKEV+28,255:POK
EV+33,0:POKEV+32,0
20 FORA=0TO62:READB:POKE832+A,B:NEXT:FOR
A=0TO7:POKE2040+A,13:NEXT
22 POKEV+37,5:POKEV+38,2:FORA=3TO10:POKE
V+39+A-2,A:NEXT
25 POKEV+1,50:POKEV+5,100:POKEV+9,150:PO
KEV+13,200
30 FORX=0TO255:POKEV,X:POKEV+4,X:POKEV+8
,X:POKEV+12,X
40 POKEV+2,50:POKEV+6,100:POKEV+10,150:P
OKEV+14,200
50 POKEV+3,X:POKEV+7,X:POKEV+11,X:POKEV+
15,X
60 NEXT
70 GOTO22
100 DATA0,0,0,0,0,0,20,0,80,5,69,64,5,69
,64,1,69,0,0,68,0,3,87,0,15,255
110 DATA192,59,251,176,255,255,252,254,2
39,188,255,255,252,239,251,236,255,255
120 DATA252,63,190,240,63,255,240,15,187
,192,3,255,0,0,204,0,0,0,0
```

SECTION 1.7

HIGH RESOLUTION GRAPHICS

High resolution graphics without 'Simon's Basic' is easy on the Commodore 64. In order to do this section you must understand exactly how programmable characters work. This is because the individual dots are programmed by their respective bits.

The high resolution screen on the Commodore 64 is 320 dots across by 200 dots down. That is a total of 64,000 individually controllable dots. You can have up to 16 colours on the screen at once, but more of that later.

Firstly we need to put the 64 into high resolution mode. When we do this it makes each line of eight bits (dots) across the screen as one byte, so that if you typed POKE 1024,1 you would get a single dot, not an 'A'. Therefore, instead of 40 bytes across by 25 down (which makes 1000), we have 40 bytes across by 200 down, making a 320 bit by 200 byte screen. This occupies 8K, so we need to move the screen upwards in memory to accommodate this. The following program does just that:

```
10 A=56576:B=53248:POKEA,PEEK(A)AND254:P
OKEB+24,8:POKEB+17,PEEK(B+17)OR32
```

Screen memory now starts at 24K (24576) and the colour memory at 16K (16384). If you type in the above program and run it, you will notice that the screen is scattered with unusual shapes of random colours. Obviously we need to clear the screen and make it a single colour (white). This program takes a long time, so be patient.

```
50 FORX=0TO1000:POKE16384+A,1:NEXT
60 FORX=0TO8000:POKE24576+A,0:NEXT
```

Now we have our high resolution screen set up, we need a routine to actually plot a point on the screen. The following subroutine does just that:

```
1000 Y1=INT(Y/8):Y2=Y-Y1*8
1010 X1=INT(X/8):X2=X-X1*8
1020 CH=(Y1*320)+(X1*8)+Y2:BI=2↑(7-X2)
1030 POKE24576+CH,PEEK(24576)ORBI
1040 RETURN
```

The above routine depends on another routine to provide the X and Y values to this routine.

The following program combines all done so far with a routine to provide X and Y values for drawing a circle:

```
10 A=56576:V=53248
20 POKEA,PEEK(A0)AND254
30 POKEV+24,8
40 POKEV+17,PEEK(V+17)OR32
50 FORX=0TO1000:POKE16384+X,1:NEXT
60 FORX=0TO8000:POKE24576+X,0:NEXT
70 C=110:D=100
80 FORS=0TO6.5STEP0.01:X=C*SIN(S)+160:Y=
D*COS(S)+100
90 X=INT(X+.5):Y=INT(Y+.5):GOSUB1000
100 NEXT
1000 Y1=INT(Y/8):Y2=Y-Y1*8
1010 X1=INT(X/8):X2=X-X1*8
1020 CH=(Y1*320)+(X1*8)+Y2:BI=2↑(7-X2)
1030 POKE24576+CH,PEEK(24576)ORBI
1040 RETURN
```

Of course, the routine in lines 70 to 100 can be changed to whatever you want.

To draw a vertical line change lines 70 to 100:

```
70 D=100
80 FORS=0TO3STEP0.01:X=160:Y=S*D
90 X=INT(X+.5):Y=INT(Y+.5):GOSUB1000
100 NEXT
```

Or to draw a diagonal line:

```
70 C=160:D=100
80 FORS=0TO2.23STEP0.01:X=S*C:Y=S*D
90 X=INT(X+.5):Y=INT(Y+.5):GOSUB1000
100 NEXT
```

Finally, the following program draws a sine curve:

```
10 A=56576:V=53248
20 POKEA,PEEK(A)AND254
30 POKEV+24,8
40 POKEV+17,PEEK(V+17)OR32
50 FORX=0TO1000:POKE16384+X,1:NEXT
60 FORX=0TO8000:POKE24576+X,0:NEXT
62 C=160:D=100
63 FORS=0TO1.999STEP0.01:X=S*C:Y=100
64 X=INT(X+.5):Y=INT(Y+.5):GOSUB1000
65 NEXT
66 D=100
67 FORS=0TO2STEP0.01:X=30:Y=S*D
68 X=INT(X+.5):Y=INT(Y+.5):GOSUB1000
69 NEXT
70 C=160:D=100
80 FORS=3TO12STEP0.01:X=S*C/5-65:Y=SIN(S
)*D/1.3+87
90 X=INT(X+.5):Y=INT(Y+.5):GOSUB1000
100 NEXT
200 STOP
1000 Y1=INT(Y/8):Y2=Y-Y1*8
1010 X1=INT(X/8):X2=X-X1*8
1020 CH=(Y1*320)+(X1*8)+Y2:BI=2↑(7-X2)
1030 POKE24576+CH,PEEK(24576+CH)ORBI
1040 RETURN
```

It would be pretty useless if we could only draw lines etc. in one colour, on one colour of background. The addresses 16384 to 17383 in the previous example control the colour of the lines and the background. The background in each 8 by 8 dot square can be controlled as well as the line or foreground colour. The value to poke for the colours are calculated by the following formula:

NUMBER = (16*foreground colour) + background colour.

The colours are the same as for the characters.

e.g. To colour the background black and the line white POKE the addresses with 16......

```
FOR A=16384 TO 17383 : POKE A, 16 :NEXT
```

Multicolour is controlled the same as it is in BASIC......

```
POKE 53265,PEEK(53265) OR 16
```

Every dot you plot will be two dots in width. Just follow the same rules as you would in multicolour programmable characters.

SECTION 1.8

OTHER COMPUTERS' BASICS

This section is for people who like the look of programs in magazines and want to type them in but find that they are not for their machine. I will cover conversion of programs in ZX81, ZX Spectrum, VIC 2Ø and Commodore Pet BASIC to Commodore 64 BASIC.

For screen displays you need to take account of screen size and adjust the variables accordingly. ZX81 and Spectrum programs contain frequent PRINT AT X,Y; statements. This command contrary to popular belief is possible on the Commodore 64. If a program has PRINT AT 21,2;"HELLO" (ZX computers) then enter the following on the 64:

```
POKE 781,21:POKE 782,2:SYS 6552Ø:PRINT"HELLO"
```

781 is the location to POKE for the column and 782 for the row. SYS6552Ø is an inbuilt routine for positioning the cursor.

If you do not adjust the scaling of the variables the program will just use part of the screen.

The following table contains various pieces of information about the four machines listed above, and what they are on the Commodore 64.

FUNCTION/ MACHINE	ZX81	SPECTRUM	VIC20	PET	COMMODORE 64
Screen locations		non-standard, but 16384 to 22528 pixels approximately.	7680 – 8185 (4096 – 4605 with 8K expansion)	32768 – 33767	1024 – 2023
Size of screen	32*22	32*22	22*23	40*25	40*25
Colour screen memory	—	Uses PRINT AT X,Y; INK colour	38400 – 38906 (37888 – 38393 with 8K expansion)	—	55296 – 56295
BASIC memory	16524 – (32768 if 16K)	22528 approximately	4096 – 7679 but anything from 1024 – 32767	1024 –	2048 – 40959 (for BASIC)
Start for programmable characters	—	Uses POKE USR"A" to POKE USR"Q"	4096/ 5120/ 6144/ 7168	—	2048 – (standard 8192)

SECTION 1.9

THE PERIPHERALS

This section is divided into four subsections:

1. Cassette file handling
2. Joysticks and paddles
3. Printer
4. Disc Drive

CASSETTE FILE HANDLING

The cassette can be used to store data files (streams of data). It does this in a serial fashion, i.e. it puts data onto the tape in a 'one after the other' way.

In order to write a file to the tape we need to open a channel to the tape. This is done with the OPEN command:

OPEN 1,1,2,"name" (or a string)

The second number of this command *must* be a one as this is the device number for the cassette recorder.

To put data onto the tape the PRINT #1 command is used. Usually this is done in a loop.

The following program copies the screen onto tape. As can be seen from the program you *must* close the file when you are finished. This is done with CLOSE 1.

```
10 OPEN1,1,2,"SCREEN SAVE"
20 FORA=1024TO2023:B=PEEK(A):C=PEEK(A+54
272)
30 PRINT#1,B:PRINT#1,C:NEXT
40 CLOSE1
```

However, the above routine is pointless unless we can retrieve the information from the tape. The INPUT #1 and the GET #1 commands do this. Their function is basically the same, but GET #1 is used where one character is wanted at a time.

To get the saved screen off the tape (saved with the previous program) use the following program:

```
10 OPEN1,1,0,"SCREEN SAVE"
20 FORA=1024TO2023:INPUT#1,B:INPUT#1,C
30 POKEA,B:POKEA+54272,C:NEXT
40 CLOSE1
```

As the cassette has to 'run through' to find data, it is not really practicable to store a lot of data onto the tape, but if speed is not required then it is ideal if used with a C90 cassette. The above method works for any data. You can use any variable or characters in quotes.

JOYSTICKS

You can use two joysticks on your Commodore 64. They plug into the ports on the right-hand side of your computer. To read the joysticks (find out their position) you need to PEEK certain addresses. I will start with Joystick 0 which plugs into the port nearest the front of the computer.

To read the joystick you PEEK address 56321. These are the values obtained......

		Sets bit X to 0
normal (no movement)	255	no bits
left	251	three
right	247	four
up	254	one
down	253	two
fire	239	five
up/left	250	one and three
up/right	246	one and four
down/left	249	two and three
down/right	245	two and four

However, if the fire button is pressed when the joystick is not at 'normal' position the above values will not hold. So the result should be ANDed. For example, to check if fire has been pressed:

IF PEEK(56321)AND 239 = 239 THEN GOSUB fire routine.

IF PEEK(56321)AND 16 = 0 THEN GOSUB fire routine.

Now for Joystick 1. To read this joystick (which plugs into the port at the back of the computer) PEEK ADDRESS 56320.

The values obtained when address 56320 is PEEKed are as follows:

		Sets bit X to 0
normal (no movement)	127	no bits
left	123	three
right	119	four
up	126	one
down	125	two
fire	111	five
up/right	118	one and three
up/left	122	one and four
down/left	117	two and three
down/right	121	two and four

As with joystick 0 the value should be ANDed to obtain a true value. For example, to move left:

IF PEEK (56320) AND 123 = 123 THEN GOSUB left routine.

This is because the AND command effectively filters out any outside values.

THE PRINTER

This section will deal with the basics of using a printer. This is mainly for people who have access to one, and for people who own a printer but don't want or need to know how to do amazing things with it. The following information works on the VIC 1515 printer and the Seikosha GP 100 VC printer, but it may well work with others.

The printer has a switch on the back that has 4 5 T on it. If you put this switch to 'T' and switch on the printer, the printer will self test. If you see any malformed characters, check the paper feed and the ribbon. If the problem still persists your printer may need attention. The positions '4' and '5' are the device number for the printer. This means that you can have two printers connected to your Commodore 64. The normal number for this switch is '4' and it is at this position that all of the following programs will work.

If you have written a program and you want a listing (copy on paper) type:

```
OPEN 4,4 : CMD 4 : LIST
```

You can use any of the formats for LIST, i.e. LIST 100–200 as if it was on the screen. When the listing is complete you need to CLOSE the channel to the printer. There is a special way to do this:

```
PRINT #4: CLOSE 4
```

If you do not do it this way the channel will not be properly closed and could cause problems later.

You can print directly onto a printer or under program control. I will do it in a program as if this is not done, the word READY is printed when the printer is finished.

Instead of printing onto the printer by using the CMD command (which sends *all* output to the printer you can selectively choose whether you want the information on the screen or the printer. This command is PRINT #4.

There are various control characters which change the output from the printer. They are as follows:

CHR$ (10) .... Line feed after printing (carriage return).

CHR$ (13) .... Line feed after printing (carriage return).

CHR$ (14)	....	Double width characters.
CHR$ (8)	....	Graphic mode command.
CHR$ (15)	....	Normal width characters.
CHR$ (16)	....	PRINT start position addressing.
CHR$ (27)	....	When followed by CHR$ (16), it is used to specify a start position according to the dot address.
CHR$ (26)	....	Repeat graphic select command.
CHR$ (145)	....	Cursor up mode command.
CHR$ (17)	....	Cursor down mode command.
CHR$ (18)	....	Reverse field characters.
CHR$ (146)	....	Reverse field off (normal characters).

For example, to print in double width characters type one of the following:

```
OPEN 4,4:CMD 4 :PRINT CHR$ (14)
OPEN 4,4:PRINT #4,CHR$ (14)
```

THE DISC DRIVE

This is a short section that lets you use VIC 1540 disc drive with the Commodore 64. This drive is not normally useable with the 64 but it can be done.

Before LOADing, SAVEing or VERIFYing, the following command *must* be typed:

```
POKE 53265,11
```

For example, to load the disc directory:

```
POKE 53265,11:LOAD "$" , 8
```

As this blanks the screen you need to bring the screen back. This is done by either pressing RUN/STOP and RESTORE or typing POKE 53265,27.

To format a disc (essential before using a disc for the first time) type:

OPEN 15,8,15 : PRINT #15, "NEW :disc name,1Ø"

Then type:

PRINT #15,"INITIALISE 1Ø,disc name,1Ø"

To save a program, type:

POKE 53265,11 :SAVE " Ø:program name, number between Ø and 255", 8 :POKE 53265,27

To verify the program, type:

POKE 53265,11:VERIFY"Ø:program name",8:POKE 53265,27

To load the program back in, type:

POKE 53265,11:LOAD"Ø:name",8

Then when the red light on the drive goes out, type POKE 53265,27 or press RUN/STOP and RESTORE.

If you have a 1541 disc drive, the commands are exactly the same except that you miss out the POKE 53265,11 and POKE 53265,27 commands.

SECTION 1.10

SPEEDING UP BASIC

Even though the BASIC on the Commodore 64 is fast for BASIC it is still relatively slow. This section will tell you how you can speed up your BASIC programs to their limit.

1. If you use variables instead of lots of numbers you save a considerable amount of time. Each variable used saves approx five to ten milliseconds. This may not sound a lot but if there are lots of loops where numbers are constantly handled, as in games or mathematical programs, the time saving is considerable.

2. When FOR NEXT loops do not put a variable after the NEXT. This also saves quite a lot of time.

3. Put all frequently used subroutines at the beginning of a program. This is because BASIC starts at the beginning of a program to look for a routine. So, if your subroutine is at the end of your program, every time you use your subroutine it has to wait until BASIC gets to it. If it is at the beginning then BASIC hardly has to look and so saves time.

4. Use FOR NEXT loops where possible in preference to loop like this:

```
10 A = A + 1:IF A< 21 THEN 10
```

5. Use single letter variables instead of double or more.

6. Try to use low line numbers (10 is better than 10,000).

Finally, if you want really fast programs use machine code.

SECTION 2.1

MACHINE CODE

Now that you have passed BASIC and into the realms of machine code, you may be asking "Well, what is machine code?". Quite simply it is the language that the microprocessor understands. In the Commodore 64 the microprocessor is a 6510 which is an upgraded 6502. Therefore, the 64 runs 6502 machine code.

Before we start, let us look at a program in BASIC and machine code:

```
10 A=1:B=10
20 C=A+B
30 PRINTC
```

That is pretty straightforward. Now look at its machine code counterpart.

```
.:1C00 A9 01 69 0A 8D 00 04 A9
.:1C08 01 8D 00 D8 60 FF FF FF
```

Well, this is pretty unintelligable isn't it. As this is so unreadable there is such a thing as assembly language. This is really just the human version of machine code. This assembly language allows us to write relatively readable programs but that run at machine code speed. Here is the above program in assembly language:

```
1C00 A9 01          LDA #$01
1C02 69 0A          ADC #$0A
1C04 8D 00 04       STA $0400
1C07 A9 01          LDA #$01
```

```
1C09 8D 00 D8      STA $D800
1C0C 60            RTS
```

This, though strange at first, is far easier to read and understand than the previous machine code program.

This section will teach assembly language and will finish by explaining how assembly language can be directly changed into machine code for entering into the 64.

As in section 1 (BASIC), I will not go through the commands in alphabetical order but in the order of simplest to hardest to understand. If you don't understand it immediately then read it again and again. It will eventually become clear and you will be writing machine code without a second thought. The commands will be listed in a tabular form with all the variations of each command in one block. Good reading and good luck!

The 6510 registers

The 6510 has what is known as registers which do all the work. They can be thought of as a kind of memory location which controls the running of any program.

Accumulator

The accumulator is the most important register in the 6510. It is where all arithmetic functions are done. It is an eight bit register and therefore can hold only a value between zero and 255.

X Register (or X Index)

This is one of the secondary registers in the 6510. It is also an eight bit register and can only hold a number up to 255. It does not allow arithmetic functions but its contents can be transferred to and from the accumulator.

Y Register (or Y Index)

This is pretty well identical in working to the X register and can perform nearly all the same functions.

The Status Register

This register consists of eight flags; a flag is something that indicates that something has or has not occurred.

The Program Counter

This is the only 16 bit register in the 651Ø. It contains the address of the current machine language instruction being executed. It is therefore always changing.

The Stack Pointer

This register contains the location of the top empty space on the stack. The stack is used for temporary storage by machine language programs and by the computer's operating system.

NUMBERING SYSTEMS

Hexadecimal numbers

Hexadecimal is the normal numeric system used in machine code. Unlike the decimal system it is not in base 1Ø but in base 16. If you find it hard to grasp then keep trying as it makes machine code programming a lot easier.

In decimal the base is 1Ø therefore there are 1Ø different digits. These are the numbers zero to nine. In hexadecimal the base is 16 and as there are only 1Ø numerals we need six more. The first six letters of the alphabet are used for this (A to F).

DECIMAL	HEXADECIMAL
Ø	ØØ
1	Ø1
2	Ø2
3	Ø3
4	Ø4
5	Ø5
6	Ø6
7	Ø7
8	Ø8
9	Ø9

10	0A
11	0B
12	0C
13	0D
14	0E
15	0F
16	10

and so on....

In decimal as you move left the powers of 10 increase by one each time. For example, 4096 is (4 *1000) + (0 * 100) + (9 * 10) + 6. In hexadecimal (Hex) it is the powers of 16 that increase by one each time. For example, 1ED2 is (1*4096) + (14 * 256) + (13* 16) + (2). In decimal the range of addressable locations is 0 to 65535, so in Hex they are 0000 to FFFF.

To distinguish these numbers from base 10 the Hex numbers are usually prefixed with a dollar sign ($). For example, 7680 in decimal is $ 1E00 in Hex.

Addressing Modes

There are various addressing modes that the 6510 uses. They will be explained now to avoid confusion later.

Absolute Addressing

This is the mode where a number is taken from a memory location for use by the processor.

eg.

```
LDA $1000
```

Immediate Addressing

This is the mode in which you load a number directly into a register.

eg.

```
LDA #$10
```

The hash (#) sign tells the assembler (program which converts assembly language into machine code) that this is a direct command.

Zero Page Addressing

This is an addressing mode that allows you to address locations Ø to 255 (Ø to F) only. The reason for this is speed. It is faster to go to a one byte number than a two byte number.

Indexed Addressing

This is the most useful mode in the 651Ø. It allows the addition of a number to a base address to go to higher addresses, i.e. 4Ø96 + 25Ø. It always involves the accumulator and the X or Y register.

eg.
LDA $ 1EØØ,X (if X contained 12 then the address would be 1E12)

From now on I will work in Hex. I will refer to the accumulator as A, the X register as X and the Y register as Y.

In order to enter the programs you will need an assembler *or* use the hexadecimal loader at the back of the book. You type the Hex numbers which are beside most of the programs into this loader and run/stop restore it to stop. Then type SYS (the starting address) to run your program.

Machine Code Instructions

The instructions will be covered in the following order:

LDA	Load A with memory or number.
LDX	Load X with memory or number.
LDY	Load Y with memory or number.
STA	Store number in memory location (from A).
STX	Store number in memory location (from X).
STY	Store number in memory location (from Y).
TAY	Transfer contents of A into Y.

TAX	Transfer contents of A into X.
TYA	Transfer contents of Y into A.
TXA	Transfer contents of X into A.
NOP	No operation (used to reserve memory).
JMP	Jump to address.
JSR	Jump to subroutine at address.
RTS	Return from subroutine.
INC	Increment (add one to) memory
INX	Increment value in X.
INY	Increment value in Y.
DEC	Decrement (subtract one from) memory.
DEX	Decrement value in X.
DEY	Decrement value in Y.
CMP	Compare A with memory/number.
CPX	Compare X with memory/number.
CPY	Compare Y with memory/number.
BEQ	Branch if value equal to.
BNE	Branch if value not equal to.
BCC	Branch if carry clear (less than).
BCS	Branch if carry set (greater than).
BVC	Branch if overflow.
BVS	Branch if no overflow.
BPL	Branch on plus.
BMI	Branch on minus.
BRK	Break (stop).
PHA	Put value in A onto stack.
PHP	Put processor statut onto stack.
PLA	Take top of stack and put in A.
PLP	Take processor status off stack.
TXS	Transfer X to stack pointer.
TSX	Transfer stack pointer to X.
AND	AND A with memory.
ORA	OR A with memory.
EOR	Exclusive OR A with memory.
BIT	Test bits with memory.

ADC	Add memory to A with carry.
SBC	Subtract memory from A with carry.
SEC	Set carry bit.
CLC	Clear carry bit.
SED	Set decimal bit.
CLD	Clear decimal bit.
SEI	Set interrupt disable bit.
CLI	Clear interrupt disable bit.
RTI	Return from interrupt routine.
CLV	Clear overflow bit.
ROR	Rotate memory one bit right.
ROL	Rotate memory one bit left.
ASL	Shift memory one bit left.
LSR	Shift memory right one bit.

COMMAND

HEX/DECIMAL

LDA #$ØØ A9	 169	Loads the accumulator with hexademical ØØ.
LDA $ØØ A5	 165	Loads the accumulator from zero page location ØØ (maximum FF).
LDA $ØØ,X B5	 181	Loads the accumulator from zero page location (ØØ + X) (maximum FF). If X contained FØ then it would load from location ØØ + FØ = FØ.
LDA $ØØØØ AD	 173	Loads accumulator from location ØØØØ (maximum FFFF).
LDA $ØØØØ,X BD	 189	Loads A from location (ØØØØ + X) (maximum FFFF). If X contained FF then it would load from location ØØØØ + FF = ØØFF.
LDA $ØØØØ,Y B9	 185	Loads A from location (ØØØØ + Y) (maximum FFFF). If Y contained DE then it would load

from location ØØØØ + DE = ØØDE.

STA $ØØ 85	133	Stores value in A in location ØØ (maximum FF).
STA $ØØ,X 95	149	Stores value in A in location (ØØ + X) (maximum FF).
STA $ØØØØ 8D	141	Stores the value of A in location ØØØØ (maximum FFFF).
STA $ØØØØ,X 9D	157	Stores the value in A in location (ØØØØ + X) (maximum FFFF).
STA $ØØØØ,Y 99	153	Stores the value of A in location (ØØØØ + Y) (maximum FFFF).
STX $ØØ 86	134	Stores the value in X in location ØØ (maximum FF).
STX $ØØ,Y 96	15Ø	Stores the value in X in location (ØØ + Y) (maximum FF).
STX $ØØØØ 8E	142	Stores the value in X in location ØØØØ (maximum FFFF).
STY $ØØ 84	132	Stores the value in Y in location ØØ (maximum FF).
LDX #$ØØ A2	162	Loads X with hexadecimal ØØ.
LDX $ØØ A6	166	Loads X with value in Hexadecimal ØØ. (maximum FF).
LDX $ØØ,Y B6	182	Loads X with value in ØØ + Y (maximum FF).
LDX $ØØØØ AE	174	Loads X with value in ØØØØ (maximum FFFF).

LDX $0000,Y BE 190	Loads X with value in 0000 + Y.
LDY #$00 A0 160	Load Y with hexadecimal 00.
LDY $00 A4 164	Load Y with value in 00 (maximum FF).
LDY $00,X B4 180	Load Y with value in location 00 + X (maximum FF).
LDY $0000 AC 172	Load Y with value in 0000 (maximum FFFF).
LDY $0000,Y BC 188	Load Y with value in 0000 + X (maximum FFFF).
STY $00,X 94 148	Stores the value in Y in location (00 + X) (maximum FF).
STY $0000 8C 140	Stores the value in Y in location 0000 (maximum FFFF).
TAY A8 168	Transfers the contents of A into Y.
TAX AA 170	Transfers the contents of A into X.
TYA 98 152	Transfers the contents of Y into A.
TXA 8A 138	Transfers the contents of X into A.
NOP EA 234	No operation. This command is the equivalent command to REM in BASIC in that it is passed by the processor. It is used to

	reserve spaces in memory in case you want to insert instructions later.
JMP $ØØØØ AC 76	Jumps to location ØØØØ (maximum FFFF). Like GOTO in BASIC.
JSR $ØØØØ 2Ø 32	Jumps to a subroutine beginning at location ØØØØ (maximum FFFF) and remembers where you came from. Like GOSUB in BASIC.
RTS 6Ø 96	Returns from a subroutine. Similar to return in BASIC.
INC $ØØ E6 23Ø	Increments (adds one to) the value in location ØØ (maximum FF). If location ØØ contained 47 it would contain 48 after this command.
INC $ØØ,X F6 246	Increments the value contained in location (ØØ+X) (maximum FF). If X contained ØA then locaiton ØØ+ØA would be incremented.
INC $ØØØØ EE 238	Increments the value contained in location ØØØØ (maximum FFFF).
INC $ØØØØ,X FE 254	Increments the value contained in location (ØØØØ+X) (maximum FFFF).
INX E8 132	Increments the value contained in the X register.
INY C8 2ØØ	Increments the value contained in the Y register.
DEC $ØØ C6 198	Decrements (subtract by one) the value in location ØØ (maximum FF).
DEC $ØØ,X D6 214	Decrements the value contained (ØØ+X) (maximum FF).

DEC $0000 CE 206	Decrements the value in 0000 (maximum FFFF).
CMP #$00 C9 201	Compares the value in 00 with the value in A (maximum FF). For example IF A=1 in BASIC.
CMP $00 C5 197	Compares the value in 00 with the value in A (maximum FF).
CMP $00,X D5 213	Compares the value in (00+X) with the value in A (maximum FF).
CMP $0000 CD 205	Compares the value in 0000 with the value in A (maximum FFFF).
CMP $0000,X DD 221	Compares the value in (0000+X) with the value in A (maximum FFFF).
CMP $0000,Y D9 217	Compares the value in (0000+Y) with the value in A (maximum FFFF).
CPX #$00 E0 224	Compares the value 00 with the value in X (maximum FF).
CPX $00 E4 228	Compares the value in 00 with the value in X (maximum FF).
CPX $0000 EC 236	Compares the value in 0000 with the value in X (maximum FFFF).
CPY #$00 C0 192	Compares the value 00 with the value in Y (maximum FF).
CPY $00 C4 196	Compares the value in 00 with the value in Y (maximum FF).
CPY $0000 CC 204	Compares the value in 0000 with the value in Y (maximum FFFF).

BEQ $1C1B F0 240	If the value compares (above) is equal to the value then branch (ie GOTO address). You can jump 127 bytes forward and 128 bytes backward. So, in the example, the command is at address 1C6B and it is told to branch back to 1C1B which is 101 (dec) backwards). The number after the 'S' sign has to be converted into a one byte decimal number by subtracting 1C1B from 1C6B (after converting to decimal).
BNE $1C1B D0 208	As above except that if value is *not* equal then branch forward/back.
BCC $1C1B 90 144	Branches forward/backward if carry clear, ie if value is less than one checked for. Has same format as BEQ.
BCS $1C1B B0 176	Branches forward/backward on carry set, ie if value is more then one checked for. Has same format as BEQ.
BVC $1C1B 50 80	Branches forward/backward if the overflow flag(V) is clear (V=0).
BVS $1C1B 70 112	Branches forward/backward if the overflow flag is set (V=1).
BPL $1C1B 10 16	Branches forward/backward if the N Flag (N) is clear (N=0), ie result is plus.
BMI $1C1B 30 48	Branches forward/backward if the N flag is set (N=1), ie result is minus.
BRK 00 0	Interrupt program. If returning to BASIC then the screen will clear and change to blue with a light blue border and characters. Same as after run/stop restore keys.

PHA	48	72	Stores the value in A on top of the stack (storage pile 256 bytes long with 'last in-first out' working).
PLA	68	104	Takes the value off top of stack and puts it into A.
PHP	08	8	Stores the processor status at top of the stack.
PLP	28	140	Takes the processor status off top of the stack.
TSX	BA	186	Transfers the stack pointer to X.
TXS	9A	154	Transfers the value in X to the stack pointer.
AND #$00	25	41	Performs a logical AND (same as in BASIC) with the value in A and Hex 00 and puts a new value in A.
AND $00	25	37	Performs a logical AND with the value in A and the value in location 00 (maximum FF) and puts a new value in A.
AND $00,X	35	53	Performs a logical AND with the value in A and location 00 + X (maximum FF) and puts a new value in A.
AND $0000	2D	45	Performs a logical AND with the value in A and location 0000 (maximum FFFF) and puts a new value in A.
AND $0000,X	3D	61	Performs a logical AND with the value in A and location 0000 + X (maximum FFFF) and puts a new value in A.

AND $0000,Y 39 57		Performs a logical AND with the value in A and location 0000+Y (maximum FFFF) and puts a new value in A.
ORA #$00 09 9		Performs a logical OR (same as in BASIC) with the value in A and Hex 00 and puts a new value in A.
ORA $00 05 5		Performs a logical OR with the value in A and the value in location 00 (maximum FF) and puts a new value in A.
ORA $00,X 15 21		Performs a logical OR with the value in A and the value in location 00+X (maximum FF) and puts a new value in A.
ORA $0000 0D 13		Performs a logical OR with the value in A and the value in location 0000 (maximum FFFF) and puts a new value in A.
ORA $0000,X 1D 29		Performs a logical OR with the value in A and the value in location 0000+X (maximum FFFF) and puts a new value in A.
ORA $0000,Y 19 25		Performs a logical OR with the value in A and the value in location 0000+Y (maximum FFFF) and puts a new value in A.
EOR #$00 49 73		Performs an exclusive OR with the value in A and Hex 00 and stores the new value in A. An exclusive OR is similar to an OR except that if and only if *one* of the binary numbers has a one then a one will be passed on (not both). Value 1 10010011 Value 2 11100111 Final 01110100 = 116 (decimal)
EOR $00		Performs an exclusive OR with the value in A

45	69	and the value in location ØØ Hex and puts a new value in A (maximum FF).
EOR $ØØ,X		Performs an exclusive OR with the value in A
55	85	and the value in location ØØ+X and puts a new value in A (maximum FF).
EOR $ØØØØ		Performs an exclusive OR with the value in A
4D	77	and the value in location ØØØØ (Hex) and puts a new value in A (maximum FFFF).
EOR $ØØØØ,X		Performs an exclusive OR with the value in A
5D	93	and the value in location ØØØØ+X and puts a new value in A (maximum FFFF).
EOR $ØØØØ,Y		Performs an exclusive OR on the value in A
59	89	and the value in location ØØØØ+Y and puts the new value in A (maximum FFFF).
BIT $ØØ		Tests (by performing a logical AND) the value
24	36	in A with the value in location ØØ(hex) (maximum FF) but leaves the value in A intact. Changes the values in flags. If A AND memory=Ø then the Z flag is set to one otherwise zero.
BIT $ØØØØ		Tests the value in A with the value in ØØØØ
2C	44	(Hex) (maximum FFFF) and adjusts the flags accordingly (as above).
ADC #$ØØ		Takes the value in A and adds it to Hex ØØ and
69	1Ø5	puts the new value in A. (If the value exceeds 255 (FF) then A will contain the value − 255.
ADC $ØØ		Takes the value in A and adds it to the value in
65	1Ø1	location ØØ(Hex) (maximum FF) and puts the new value in A.
ADC $ØØ,X		Takes the value in A and adds it to the value in

75	117	Hex 00+X (maximum FF) and puts the new value in A.
ADC $0000 6D	 109	Takes the value in A and adds it to the value in location 0000(maximum FFFF) and puts new value in A.
ADC $0000,X 7D	 125	Takes the value in A and adds it to the value in location 0000+X (maximum FFFF) and puts the new value in A.
ADC $0000,Y 79	 121	Takes the value in A and adds it to the value in Hex 0000+Y (maximum FFFF) and puts the new value in A.
SBC #$00 E9	 233	Takes the value in A and subtracts Hex 00 from it and puts the new value in A.
SBC $00 E5	 229	Takes the value in A and subtracts the value in location 00 from it (maximum FF) and puts the new value in A.
SBC $00,X F5	 245	Takes the value in A and subtracts the value in location 00+X (maximum FF) from it and puts new value in A.
SBC $0000 ED	 237	Takes the value in A and subtracts the value in location 0000 (maximum FFFF) from it and puts the new value in A.
SBC $0000,X FD	 253	Take the value in A and subtracts the value in location 0000+X (maximum FFFF) from it and puts the new value in A.
SBC $0000,Y F9	 249	Takes the value in A and subtracts the value in location 0000+Y (maximum FFFF) from it and puts the new value in A.

SEC 38	56	Sets the carry flag, ie puts the carry flag to one (C = 1).
CLC		Clears the carry flag, ie puts the carry flag to value zero (C = Ø).
SED F8	248	Sets the decimal mode, ie puts the D flag equal to one (D = 1).
CLD D8	216	Clears the decimal mode, ie puts the D flag equal to zero (D = Ø).
SEI 78	12Ø	Sets the interrupt disable status and turns off all unnecessary interrupts (I = 1) — see the interrupts section.
CLI 58	88	Clears the interrupt disable status, ie turns of all the interrupts again — see the interrupts section.
RTI 4Ø	64	Returns from an interrupt. Used instead of RTS when returning from an interrupt.
CLV B8	184	Clears the overflow flag (V = Ø).
ROR 6A	1Ø6	Rotates the bits in the accumulator right by one bit: C 7 6 5 4 3 2 1 Ø So: 1ØØØØØØ1 would become: 11ØØØØØØ
ROR $ØØ 66	1Ø2	Rotates the bits of location ØØ(Hex) (maximum FF) right by one bit.
ROR $ØØ,X 76	118	Rotates the bits of location ØØ + X (maximum FF) right by one bit.

ROR $0000 6E	 110	Rotates the bits of location 0000 (maximum FFFF) right by one bit.
ROR $0000,X 7E	 126	Rotates the bits of location 0000+X (maximum FFFF) right by one bit.
ROL 2A	 42	Rotates the bits in the accumulator left by one bit: 7 6 5 4 3 2 1 0 C So: 10000001 would become: 00000011
ROL $00 26	 38	Rotates the bits of location 00(Hex) (maximum FF) left by one bit.
ROL $00,X 36	 54	Rotates the bits of location 00+X (maximum FF) left by one bit.
ROL $0000 2E	 46	Rotates the bits in location 0000 (maximum FFFF) left by one bit.
ROL $0000,X 3E	 62	Rotates the bits in location 0000+X (maximum FFFF) left by one bit.
ASL 0A	 10	Shifts the contents of A left by one bit and puts the bit knocked off onto carry and puts a zero in at the right. C 7 6 5 4 3 2 1 0 0 11000001 would go to: 10000010
ASL $00 06	 6	Shifts the contents of 00 (maximum FF) left by one bit and puts a zero on at the right.
ASL $00,X 16	 22	Shifts the contents of 00+X (maximum FF) left by one bit and puts a zero on at the right.

ASL $0000 03	14	Shift the contents of location 0000 (maximum FFFF) left by one bit and puts a zero on at the right.
ASL $0000,X 1E	30	Shifts the contents of location 0000+X left by one and puts a zero on at the right (maximum FFFF).
LSR 4A	74	Shifts the contents of A right by one bit and puts a zero on at the left and puts the bit knocked off onto carry. 0 7 6 5 4 3 2 1 0 C eg. 0 100000001 C would be: 01000000
LSR $00 46	70	Shifts the contents of location 00 right by one bit and puts a zero on at the left (maximum FF).
LSR $00,X 56	86	Shifts the contents of location 00+X (maximum FF) right by one bit and puts a zero on at the left.
LSR $0000 4E	78	Shifts the contents of location 0000 (maximum FFFF) right by one bit and puts a zero on at the left.
LSR $0000,X 5E	94	Shifts the contents of location 0000+X right by one bit and puts a zero on at the left (maximum FFFF).

CONVERTING ASSEMBLY LANGUAGE TO MACHINE CODE

As assembly language is far easier to use than machine code, if you don't have an assembler you should write assembly language programs and change them into Hex for entering with the Hex loader at the back of the book. This is how to do the conversion and how to start or stop your machine code programs.

In the table (at the beginning of section 2.1) I gave the assembly language form, and its Hex and decimal equivalent. These numbers are the value for the mnemonic (three letter word),, the operand (# $ ØØ, for example, has to be calculated. The following will deal with hex numbers. If the operand goes up to FF then it will occupy one byte so you just keep the operand to POKE directly into memory (or using the Hex loader). But if it goes up to FFFF then it is a two byte operand and has a different format. The 651Ø works in low byte/high byte order. The low byte of the two byte address is the second two Hex digits and the high byte is the first two.

eg.
To change LDA $ Ø1FE into machine code:

The value for LDA$ is AD (or 173), the second two Hex digits are FE, and the first two are Ø1. So you enter the numbers into the Hex loader in the order AD,FE,Ø1. (Or POKE into memory....173, 254, 1).

In the following sections, the programs are listed in assembly language and in a Hex dump. This is a list of the Hex numbers to enter into the loader. You only enter the first two byte number as it is the starting address, then you enter all the Hex numbers.

To start the assembly language programs type G starting address into your assembler or SYS starting address. To start the Hex dump versions type SYS starting address. For example to start either program (starting address is 7168)

SYS 7168

SECTION 2.2

COLOUR IN MACHINE CODE

In matching code there are three ways of colouring characters:

1. Put the colour value into the accumulator and transfer it to location 646 (as in BASIC).

eg.
To make characters orange:

```
1C00 A9 0B          LDA #$0B
1C02 8D 86 02       STA $0286
1C05 60             RTS
```

2. Put CHR$ (value for the colour — see Appendix 4) into the accumulator and then JSR to the PRINT subroutine at location 65490 ($FFD2_6$).

eg.
To make the characters orange:

```
1C00 A9 31          LDA #$31
1C02 20 D2 FF       JSR $FFD2
1C05 60             RTS
```

3. To colour individual characters on the screen (like POKEing colour) character by character.

eg.
To colour the top line of the screen orange:

```
1C00 A9 08        LDA #$08
1C02 A2 00        LDX #$00
1C04 9D 00 D8     STA $D800,X
1C07 E8           INX
1C08 E0 28        CPX #$28
1C0A D0 F8        BNE $1C04
1C0C 60           RTS
```

The screen and border colours are controlled in the same way as they are in BASIC, by using locations 53280 and 53281.

eg.
To change the screen colour to black:

```
1C00 A9 00        LDA #$00
1C02 8D 21 D0     STA $D021
1C05 60           RTS
```

Or to change the border to red:

```
1C00 A9 02        LDA #$02
1C02 8D 20 D0     STA $D020
1C05 60           RTS
```

SECTION 2.3

SOUND IN MACHINE CODE

Sound will not be covered in such depth as in the BASIC section as all the values and tables and POKEs are identical, and they have already been covered.

To produce sound in machine code you load the accumulator (or the X or the Y register) with the value for the waveform, note low or high frequency, attack/decay, etc. Then you store this value in the respective register.

eg.
To create a sawtooth waveform on voice 1:

```
1C00 A9 21        LDA #$21
1C02 8D 04 D4     STA $D404
1C05 60           RTS
```

In machine code sound needs a delay loop to make the sound slow enough to sound correct.

The following routine has a sound effect (a red alert) without a delay, and the second one is the same with a delay:

```
1C00 A9 21        LDA #$21
1C02 8D 04 D4     STA $D404
1C05 A9 80        LDA #$80
1C07 8D 05 D4     STA $D405
1C0A 8D 06 D4     STA $D406
1C0D A9 0A        LDA #$0A
1C0F 8D 18 D4     STA $D418
1C12 A2 E6        LDX #$E6
```

```
1C14 8D 00 D4      STA $D400
1C17 8E 01 D4      STX $D401
1C1A CA            DEX
1C1B E0 64         CPX #$64
1C1D D0 F5         BNE $1C14
1C1F 4C 12 1C      JMP $1C12
```

For those without an assembler, use the Hex loader at the back of the book and type in the following numbers (the first is the Hex starting address):

```
B*
   PC   SR AC XR YR S$
.;97FE 32 00 00 00 F6
.
.:1C00 A9 21 8D 04 D4 A9 80 8D
.:1C08 05 D4 8D 06 D4 A9 0A 8D
.:1C10 18 D4 A2 E6 8D 00 D4 8E
.:1C18 01 D4 CA E0 64 D0 F5 4C
.:1C20 12 1C 59 00 00 00 00 00
```

```
1C00 A9 21         LDA #$21
1C02 8D 04 D4      STA $D404
1C05 A9 80         LDA #$80
1C07 8D 05 D4      STA $D405
1C0A 8D 06 D4      STA $D406
1C0D A9 0A         LDA #$0A
1C0F 8D 18 D4      STA $D418
1C12 A2 E6         LDX #$E6
1C14 8E 00 D4      STX $D400
1C17 8E 01 D4      STX $D401
1C1A CA            DEX
1C1B 20 30 1C      JSR $1C30
1C1E E0 64         CPX #$64
1C20 D0 F5         BNE $1C17
1C22 A9 00         LDA #$00
1C24 8D 00 D4      STA $D400
1C27 8D 01 D4      STA $D401
```

```
1C2A 20 30 1C      JSR $1C30
1C2D 4C 12 1C      JMP $1C12
1C30 A0 00         LDY #$00
1C32 C8            INY
1C33 D0 FD         BNE $1C32
1C35 60            RTS
```

And here are the Hex numbers:

```
B*
   PC  SR AC XR YR S$
.;97FE 32 00 00 00 F6
.
.:1C00 A9 21 8D 04 D4 A9 80 8D
.:1C08 05 D4 8D 06 D4 A9 0A 8D
.:1C10 18 D4 A2 E6 8E 00 D4 8E
.:1C18 01 D4 CA 20 30 1C E0 64
.:1C20 D0 F5 A9 00 8D 00 D4 8D
.:1C28 01 D4 20 30 1C 4C 12 1C
.:1C30 A0 00 C8 D0 FD 60 50 31
```

The above routine forms the basis for all sound effects in machine code.

To play tunes, instead of using data statements, you POKE the notes into a spare part of memory and LDA $ value, X for the tune.

The following program uses this method to play a scale. The first part of the program is in BASIC and it POKEs the note values into memory.

```
10 FORA=0TO16:READB:POKE832+A,B:NEXT
20 SYS7168
30 DATA 75,34,126,38,52,43,198,45,97,51,
172,57,188,64,149,68,0,0
```

```
B*
   PC   SR AC XR YR S$
.;97FE 32 00 00 00 F6
.
1C00 A9 21          LDA #$21
1C02 8D 04 D4       STA $D404
1C05 A9 80          LDA #$80
1C07 8D 05 D4       STA $D405
1C0A 8D 06 D4       STA $D406
1C0D A9 0A          LDA #$0A
1C0F 8D 18 D4       STA $D418
1C12 A2 00          LDX #$00
1C14 BD 3C 03       LDA $033C,X
1C17 8D 00 D4       STA $D400
1C1A BD 3D 03       LDA $033D,X
1C1D 8D 01 D4       STA $D401
1C20 20 2C 1C       JSR $1C2C
1C23 E8             INX
1C24 E8             INX
1C25 E0 14          CPX #$14
1C27 D0 EB          BNE $1C14
1C29 4C 12 1C       JMP $1C12
1C2C 8A             TXA
1C2D 48             PHA
1C2E A2 00          LDX #$00
1C30 A0 00          LDY #$00
1C32 C8             INY
1C33 D0 FD          BNE $1C32
1C35 E8             INX
1C36 D0 FA          BNE $1C32
1C38 68             PLA
1C39 AA             TAX
1C3A 60             RTS
```

And here is the Hex dump for the program:

```
.:1C00 A9 21 8D 04 D4 A9 80 8D
.:1C08 05 D4 8D 06 D4 A9 0A 8D
.:1C10 18 D4 A2 00 BD 3C 03 8D
```

```
.:1C18 00 D4 BD 3D 03 8D 01 D4
.:1C20 20 2C 1C E8 E8 E0 14 D0
.:1C28 EB 4C 12 1C 8A 48 A2 00
.:1C30 A0 00 C8 D0 FD E8 D0 FA
.:1C38 68 AA 60 41 00 85 08 00
```

SECTION 2.4

ANIMATION IN MACHINE CODE

The principles in animation are basically the same as in BASIC. When you are moving a character you must erase its previous position before moving to a new one. However, in machine code you need to slow the action down to see it. For example, to move a ball across the screen in BASIC:

```
5 X=0
10 POKE1024+X,81:POKE55296+X,1:X=X+1:POK
E1023+X,32:IFX=40THENEND
20 GOTO10
```

Would be the following in machine code:

```
1C00 A2 00          LDX #$00
1C02 A9 51          LDA #$51
1C04 9D 00 04       STA $0400,X
1C07 A9 01          LDA #$01
1C09 9D 00 D8       STA $D800,X
1C0C A9 20          LDA #$20
1C0E 9D FF 03       STA $03FF,X
1C11 E8             INX
1C12 E0 28          CPX #$28
1C14 D0 EC          BNE $1C02
1C16 60             RTS
```

And here is the Hex dump:

```
.:1C00 A2 00 A9 51 9D 00 04 A9
.:1C08 01 9D 00 D8 A9 20 9D FF
.:1C10 03 E8 E0 28 D0 EC 60 8D
.?
```

However, you would need to slow this routine down as it is too fast to see. You would need a routine like this:

```
.
1C17 A2 00          LDX #$00
1C19 A0 00          LDY #$00
1C1B C8             INY
1C1C D0 FD          BNE $1C1B
1C1E E8             INX
1C1F D0 FA          BNE $1C1B
1C21 60             RTS
```

And here is the hex dump:

```
.
.:1C17 A2 00 A0 00 C8 D0 FD E8
.:1C1F D0 FA 60 1C E8 E8 E0 14
```

The main routine uses the X register as well as the delay so we need to store its value on the stack thus:

```
1C17 8A             TXA
1C18 48             PHA
1C19 A2 00          LDX #$00
1C1B A0 00          LDY #$00
1C1D C8             INY
1C1E D0 FD          BNE $1C1D
1C20 E8             INX
1C21 D0 F8          BNE $1C1B
1C23 68             PLA
1C24 AA             TAX
1C25 60             RTS
```

```
.:1C17 8A 48 A2 00 A0 00 C8 D0
.:1C1F FD E8 D0 F8 68 AA 60 14
```

So the completed routine looks like this:

```
.
1C00 A2 00          LDX #$00
1C02 A9 51          LDA #$51
```

```
1C04 9D 00 04      STA $0400,X
1C07 A9 01         LDA #$01
1C09 9D 00 D8      STA $D800,X
1C0C A9 20         LDA #$20
1C0E 9D FF 03      STA $03FF,X
1C11 E8            INX
1C12 20 1A 1C      JSR $1C1A
1C15 E0 28         CPX #$28
1C17 D0 E9         BNE $1C02
1C19 60            RTS
1C1A 8A            TXA
1C1B 48            PHA
1C1C A2 00         LDX #$00
1C1E A0 00         LDY #$00
1C20 C8            INY
1C21 D0 FD         BNE $1C20
1C23 E8            INX
1C24 D0 FA         BNE $1C20
1C26 68            PLA
1C27 AA            TAX
1C28 60            RTS
```

```
.
.:1C00 A2 00 A9 51 9D 00 04 A9
.:1C08 01 9D 00 D8 A9 20 9D FF
.:1C10 03 E8 20 1A 1C E0 28 D0
.:1C18 E9 60 8A 48 A2 00 A0 00
.:1C20 C8 D0 FD E8 D0 FA 68 AA
.:1C28 60 00 00 00 00 00 00 00
```

To start type SYS START. If this seems slow for machine code remember that the computer is counting to 65536 every time it moves the ball (as a delay). Imagine that in BASIC — it would take hours.

Animation, therefore, depends on indexed (or added to) addressing.

But we now come upon a problem. If we can only add 255 to our base (STA $, for example) how do we cover the whole screen as it is 1,000 locations in length?

Well, there are two ways. One is to actually change the program as it is running; and the other is to have different loops to add up to 1,000. For the first method to move a ball through every location we would need the following routine:

```
1C00 A2 00        LDX #$00
1C02 A9 81        LDA #$81
1C04 9D 00 04     STA $0400,X
1C07 A9 20        LDA #$20
1C09 9D FF 03     STA $03FF,X
1C0C E8           INX
1C0D D0 F3        BNE $1C02
1C0F AC 06 1C     LDY $1C06
1C12 C8           INY
1C13 8D 06 1C     STA $1C06
1C16 AC 0B 1C     LDY $1C0B
1C19 C8           INY
1C1A 8C 0B 1C     STY $1C0B
1C1D C0 07        CPY #$07
1C1F D0 E1        BNE $1C02
1C21 60           RTS
```

```
.:1C00 A2 00 A9 81 9D 00 04 A9
.:1C08 20 9D FF 03 E8 D0 F3 AC
.:1C10 06 1C C8 8D 06 1C AC 0B
.:1C18 1C C8 8C 0B 1C C0 07 D0
.:1C20 E1 60 FD E8 D0 FA 68 AA
```

The second method would be:

```
1C00 A2 00        LDX #$00
1C02 A9 81        LDA #$81
1C04 9D 00 04     STA $0400,X
1C07 A9 01        LDA #$01
```

```
1C09 9D 00 D8      STA $D800,X
1C0C E8            INX
1C0D D0 F3         BNE $1C02
1C0F A9 81         LDA #$81
1C11 9D FF 04      STA $04FF,X
1C14 A9 01         LDA #$01
1C16 9D FF D8      STA $D8FF,X
1C19 E8            INX
1C1A D0 F3         BNE $1C0F
1C1C A9 81         LDA #$81
1C1E 9D FE 05      STA $05FE,X
1C21 A9 01         LDA #$01
1C23 9D FE D9      STA $D9FE,X
1C26 E8            INX
1C27 D0 F3         BNE $1C1C
1C29 A9 81         LDA #$81
1C2B 9D FD 06      STA $06FD,X
1C2E A9 01         LDA #$01
1C30 9D FD DA      STA $DAFD,X
1C33 E8            INX
1C34 D0 F3         BNE $1C29
1C36 60            RTS
```

And here is the Hex dump:

```
.:1C00 A2 00 A9 81 9D 00 04 A9
.:1C08 01 9D 00 D8 E8 D0 F3 A9
.:1C10 81 9D FF 04 A9 01 9D FF
.:1C18 D8 E8 D0 F3 A9 81 9D FE
.:1C20 05 A9 01 9D FE D9 E8 D0
.:1C28 F3 A9 81 9D FD 06 A9 01
.:1C30 9D FD DA E8 D0 F3 60 00
```

As you can see, the first method is far more efficient. However, if the program is to be stored on a ROM cartridge then the second method would have to be used.

If you want user interaction, ie movement because of a joystick or keyboard input, you would load the accumulator with the respective address (197 for keyboard) and compare this with the various numbers and then jump to the respective routine.

eg.

The following program moves a character left and right along the bottom line of the screen according to keyboard input. The 'Z' key moves the character left and the 'M' key moves it right.

```
1C00 A2 14       LDX #$14
1C02 A0 14       LDY #$14
1C04 AD C5 00    LDA $00C5
1C07 C9 24       CMP #$24
1C09 F0 05       BEQ $1C10
1C0B C9 0C       CMP #$0C
1C0D F0 09       BEQ $1C18
1C0F 60          RTS
1C10 E0 27       CPX #$27
1C12 F0 FB       BEQ $1C0F
1C14 E8          INX
1C15 4C 23 1C    JMP $1C23
1C18 E0 00       CPX #$00
1C1A F0 F3       BEQ $1C0F
1C1C CA          DEX
1C1D 4C 23 1C    JMP $1C23
1C20 4C 0F 1C    JMP $1C0F
1C23 A9 01       LDA #$01
1C25 9D C0 07    STA $07C0,X
1C28 9D C0 DB    STA $DBC0,X
1C2B CA          DEX
1C2C A9 20       LDA #$20
1C2E 9D C0 07    STA $07C0,X
1C31 E8          INX
1C32 E8          INX
1C33 9D C0 07    STA $07C0,X
1C36 CA          DEX
1C37 4C 0F 1C    JMP $1C0F
```

```
.
.:1C00 A2 14 A0 14 AD C5 00 C9
.:1C08 24 F0 05 C9 0C F0 09 60
```

```
.:1C10 E0 27 F0 FB E8 4C 23 1C
.:1C18 E0 00 F0 F3 CA 4C 23 1C
.:1C20 4C 0F 1C A9 01 9D C0 07
.:1C28 9D C0 DB CA A9 20 9D C0
.:1C30 07 E8 E8 9D C0 07 CA 4C
.:1C38 0F 1C 00 00 00 00 00 00
```

Rather than making the above program run so fast that you cannot see, it will jump to it from BASIC.

```
10 SYS7168
20 SYS7172:GOTO20
```

Line 10 sets up the registers and line 20 goes to the machine code until you press RUN/STOP.

If you follow the rules of animation in BASIC and translate them to machine code you can write machine code animation without any problems.

If your program does not work first time don't worry, few machine code programs work first time. Check it again and change anything that you see necessary, and you *will* eventually succeed.

SECTION 2.5

PROGRAMMABE CHARACTERS IN MACHINE CODE

It is not practicable to create these characters in machine code as it would require a minimum of 40 bytes to create over and above the eight bytes needed to store them. However, the occasional one could be created by the following method:

```
LDA row 1 value
STA 8192 + row1 + (character code*8)
LDA row 2 value
...
...
...
to row 7.
```

The second method would involve indexed addressing (ie adding to) as in the following example:

```
1C00 A2 00        LDX #$00
1C02 BD 00 10     LDA $1000,X
1C05 9D 00 20     STA $2000,X
1C08 E8           INX
1C09 E0 07        CPX #$07
1C0B D0 F5        BNE $1C02
1C0D 60           RTS
```

The above routine would require the data for the '@' already stored in locations 1000 to 1007 Hex.

However, using programmable characters in machine code is very useful. In animation, all you do is write a program which uses normal characters and then program the characters that you want. Have the first part of your program as follows:

```
LDA #$24
STA $D018
```

This puts the computer into the programmed character mode. Your program will still run normally and PEEKing the characters still gives the same results, but it looks much better for you.

If you want the full character set copied into RAM rather than POKE each character into RAM, you could use the following routine:

```
1C00 78          SEI
1C01 A5 01       LDA $01
1C03 29 FB       AND #$FB
1C05 85 01       STA $01
1C07 A2 00       LDX #$00
1C09 BD 00 D0    LDA $D000,X
1C0C 9D 00 20    STA $2000,X
1C0F E8          INX
1C10 D0 F7       BNE $1C09
1C12 BD FF D0    LDA $D0FF,X
1C15 9D FF 20    STA $20FF,X
1C18 E8          INX
1C19 D0 F7       BNE $1C12
1C1B A5 01       LDA $01
1C1D 09 04       ORA #$04
1C1F 85 01       STA $01
1C21 58          CLI
1C22 A9 18       LDA #$18
1C24 8D 18 D0    STA $D018
1C27 60          RTS

.:1C00 78 A5 01 29 FB 85 01 A2
.:1C08 00 BD 00 D0 9D 00 20 E8
.:1C10 D0 F7 BD FF D0 9D FF 20
.:1C18 E8 D0 F7 A5 01 09 04 85
.:1C20 01 58 A9 18 8D 18 D0 60
```

You will notice that the cursor looks odd. This is because only the first 64 characters have been re-programmed.

The first four lines and the last four lines are for paging out the ROM as it is 'hidden' behind the I/O. The first four page the character ROM in the last four page it out. (Not the RTS.)

To get multicolour characters you simply create the characters in BASIC (as you would do in BASIC) and then POKE the values of the colours as in BASIC into locations 53282 and 53283 (plus the character colour).

The following program changes the colours of the characters to red, white and orange:

```
1C00 A9 98          LDA #$98
1C02 8D 16 D0       STA $D016
1C05 A9 08          LDA #$08
1C07 8D 86 02       STA $0286
1C0A A9 02          LDA #$02
1C0C 8D 22 D0       STA $D022
1C0F A9 01          LDA #$01
1C11 8D 23 D0       STA $D023
1C14 60             RTS
```

```
.:1C00 A9 98 8D 16 D0 A9 08 8D
.:1C08 86 02 A9 02 8D 22 D0 A9
.:1C10 01 8D 23 D0 60 9D FF 20
```

This program also puts the 64 into multicolour mode.

SECTION 2.6

SPRITES IN MACHINE CODE

Using sprites in machine code is identical to using them in BASIC. It is easier to create them in BASIC but they can still be saved and used as part of a machine code program.

eg.
Instead of typing POKE V+21,4 type:

```
1C00 A9 04          LDA #$04
1C02 8D 15 D0       STA $D015
1C05 60             RTS
```

The following programs move a sprite over the bottom of the screen. The sprite will be a square block. The programs will show you that all you do to literally convert the POKEs to LDAs and STAs.

```
10 FORA=0TO62:POKE832+A,255:NEXT
20 POKE53269,4 :POKE2042,13
30 IFPEEK(197)=4THEN100
40 IFPEEK(197)=3THEN200
50 IFPEEK(197)=10THEN300
60 IFPEEK(197)=18THEN400
70 GOTO30
100 IFPEEK(53253)>46THENY=Y-1
110 POKE53253,Y
120 GOTO70
200 IFPEEK(53253)<225THENY=Y+1
210 POKE53253,Y
220 GOTO70
300 IFPEEK(53252)>30ANDPEEK(53264)=0THEN
```

```
X=X-1:GOTO310
305 IFPEEK(53264)=4ANDPEEK(53252)>0THENX
=X-1
307 IFPEEK(53264)=4ANDPEEK(53253)=0THENP
OKE53264,0:X=255
310 POKE53252,X
320 GOTO70
400 IFPEEK(53252)<255THENX=X+1
410 IFPEEK(53252)=255THENPOKE53264,4:X=0
420 IFPEEK(53264)=4ANDPEEK(53252)=70THEN
440
430 POKE53252,X
440 GOTO70
```

And in machine code.

Use the BASIC program to run this machine code. If you type RUN 20 then you will see the speed of machine code.

```
1 FORA=0TO62:POKE832+A,255:NEXT:POKE2042
,13:POKE53269,4
10 SYS7168:GOTO10
20 SYS7152
```

```
1BF0 20 00 1C      JSR $1C00
1BF3 4C F0 1B      JMP $1BF0
1BF6 EA            NOP
1BF7 EA            NOP
1BF8 EA            NOP
1BF9 EA            NOP
1BFA EA            NOP
1BFB EA            NOP
1BFC EA            NOP
1BFD 4C A9 1C      JMP $1CA9
1C00 A5 C5         LDA $C5
1C02 C9 40         CMP #$40
1C04 F0 14         BEQ $1C1A
1C06 C9 12         CMP #$12
1C08 F0 11         BEQ $1C1B
```

```
1C0A C9 0A      CMP #$0A
1C0C F0 46      BEQ $1C54
1C0E C9 04      CMP #$04
1C10 F0 7B      BEQ $1C8D
1C12 C9 03      CMP #$03
1C14 F0 E7      BEQ $1BFD
1C16 C9 0D      CMP #$0D
1C18 F0 79      BEQ $1C93
1C1A 60         RTS
1C1B A8         TAY
1C1C AD 10 D0   LDA $D010
1C1F 29 04      AND #$04
1C21 C9 04      CMP #$04
1C23 F0 0F      BEQ $1C34
1C25 AE 04 D0   LDX $D004
1C28 E0 FF      CPX #$FF
1C2A F0 17      BEQ $1C43
1C2C E8         INX
1C2D 8E 04 D0   STX $D004
1C30 98         TYA
1C31 4C 16 1C   JMP $1C16
1C34 AE 04 D0   LDX $D004
1C37 E0 3F      CPX #$3F
1C39 F0 04      BEQ $1C3F
1C3B E8         INX
1C3C 8E 04 D0   STX $D004
1C3F 98         TYA
1C40 4C 16 1C   JMP $1C16
1C43 AD 10 D0   LDA $D010
1C46 09 04      ORA #$04
1C48 8D 10 D0   STA $D010
1C4B A2 00      LDX #$00
1C4D 8E 04 D0   STX $D004
1C50 98         TYA
1C51 4C 16 1C   JMP $1C16
1C54 A8         TAY
1C55 AD 10 D0   LDA $D010
1C58 29 04      AND #$04
1C5A C9 04      CMP #$04
1C5C F0 0F      BEQ $1C6D
```

```
1C5E AE 04 D0     LDX $D004
1C61 E0 16        CPX #$16
1C63 F0 04        BEQ $1C69
1C65 CA           DEX
1C66 8E 04 D0     STX $D004
1C69 98           TYA
1C6A 4C 16 1C     JMP $1C16
1C6D AE 04 D0     LDX $D004
1C70 E0 00        CPX #$00
1C72 F0 08        BEQ $1C7C
1C74 CA           DEX
1C75 8E 04 D0     STX $D004
1C78 98           TYA
1C79 4C 16 1C     JMP $1C16
1C7C AD 10 D0     LDA $D010
1C7F 29 FB        AND #$FB
1C81 8D 10 D0     STA $D010
1C84 A2 FF        LDX #$FF
1C86 8E 04 D0     STX $D004
1C89 98           TYA
1C8A 4C 16 1C     JMP $1C16
1C8D AE 05 D0     LDX $D005
1C90 E0 2E        CPX #$2E
1C92 F0 11        BEQ $1CA5
1C94 AD 1E D0     LDA $D01E
1C97 29 04        AND #$04
1C99 C9 04        CMP #$04
1C9B D0 04        BNE $1CA1
1C9D 98           TYA
1C9E 4C 16 1C     JMP $1C16
1CA1 CA           DEX
1CA2 8E 05 D0     STX $D005
1CA5 98           TYA
1CA6 4C 16 1C     JMP $1C16
1CA9 AE 05 D0     LDX $D005
1CAC E0 ED        CPX #$ED
1CAE F0 0D        BEQ $1CBD
1CB0 AD 1E D0     LDA $D01E
1CB3 29 04        AND #$04
1CB5 C9 04        CMP #$04
```

```
1CB7 F0 04        BEQ $1CBD
1CB9 E8           INX
1CBA 8E 05 D0     STX $D005
1CBD 98           TYA
1CBE 4C 16 1C     JMP $1C16
```

```
.
.:1BF0 20 00 1C 4C F0 1B EA EA
.:1BF8 EA EA EA EA EA 4C A9 1C
.:1C00 A5 C5 C9 40 F0 14 C9 12
.:1C08 F0 11 C9 0A F0 46 C9 04
.:1C10 F0 7B C9 03 F0 E7 C9 0D
.:1C18 F0 79 60 A8 AD 10 D0 29
.:1C20 04 C9 04 F0 0F AE 04 D0
.:1C28 E0 FF F0 17 E8 8E 04 D0
.:1C30 98 4C 16 1C AE 04 D0 E0
.:1C38 3F F0 04 E8 8E 04 D0 98
.:1C40 4C 16 1C AD 10 D0 09 04
.:1C48 8D 10 D0 A2 00 8E 04 D0
.:1C50 98 4C 16 1C A8 AD 10 D0
.:1C58 29 04 C9 04 F0 0F AE 04
.:1C60 D0 E0 16 F0 04 CA 8E 04
.:1C68 D0 98 4C 16 1C AE 04 D0
.:1C70 E0 00 F0 08 CA 8E 04 D0
.:1C78 98 4C 16 1C AD 10 D0 29
.:1C80 FB 8D 10 D0 A2 FF 8E 04
.:1C88 D0 98 4C 16 1C AE 05 D0
.:1C90 E0 2E F0 11 AD 1E D0 29
.:1C98 04 C9 04 D0 04 98 4C 16
.:1CA0 1C CA 8E 05 D0 98 4C 16
.:1CA8 1C AE 05 D0 E0 ED F0 0D
.:1CB0 AD 1E D0 29 04 C9 04 F0
.:1CB8 04 E8 8E 05 D0 98 4C 16
.:1CC0 1C FF FF FF FF FF FF 7B
```

SECTION 2.7

COMMODORE 64 ARCHITECTURE AND INTERRUPTS

The memory layout in the Commodore 64 is strange to say the least. Its microprocessor (the 6510) can directly address 64K of memory but the 64 has a total of 84K of memory. How??? Well, the 6510 has a special port built in for switching in and out blocks of memory. It is controlled by address one.

The 64 has 64K of RAM. But without INPUT/OUTPUT facilities it is useless. As the full 64K is already filled we will have to put the other things on top. In the 64, the I/O is from addresses D000(53248) to DFFF (57343). This covers some RAM but we can flip the I/O out to expose the underlying RAM. The I/O consists of two CIA chips for external communication (cassette, keyboard, RS232, etc), the SID sound chip and the video chip.

Now we need an operating system, as a computer does nothing without one. In the 64 it is called the KERNAL and it is located from location E000 (57344) to FFF (65535) on top of the RAM. This can also be removed to expose the underlying RAM.

We now need a language. This language is BASIC. It is located from locations A000(40906) to BFFF (49152).

The computer will now work — or will it? Yes it will, but we will have no characters to display on the screen. Therefore, we need another chip. This is called the character ROM or the character generator. It stores the 'bit patterns' for the characters which can be displayed on the screen. It is stored from location D000 (53248) to DFFF (57343). *Wait a minute*. . . . . the I/O is stored there. Never mind, we will just put the character ROM under it. The video chip has special circuitary to enable it to use the character ROM at all times.

The picture of the memory is very strange. It looks like this:

Commodore 64 Memory

Addresses shown in hexadecimal.

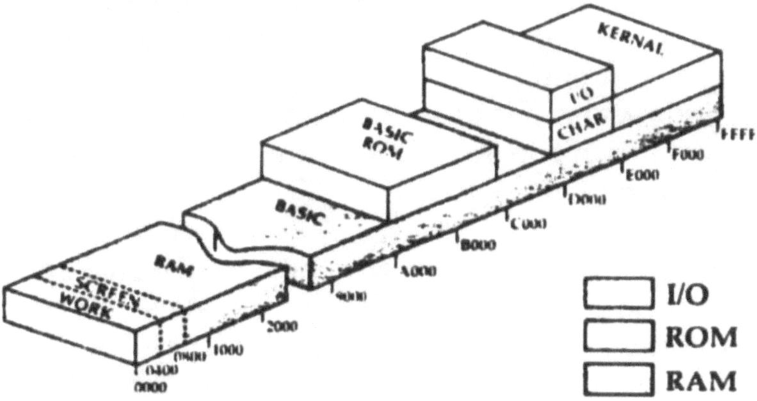

This type of memory organisation is extremely flexible, and it allows us to tailor the system to our needs.

Firstly, in the programmable characters section, we used this facility to page in the character ROM (put it on top of the pile so that it can be PEEKed):

```
1 POKE56334,128:POKE1,51:X=PEEK(53248):P
OKE1,55:POKE56334,129:PRINTX
```

Typing the above would give a result of 60 (the first value in the character ROM).

When you have paged something in you must return the computer to its normal mode by paging out what was paged in.

When paging occurs at certain places the interrupts *must* be turned off. This is only when you page out the operating system (KERNAL) or the I/O. This is done with POKE 56334,128 (or SEI in machine code). When you have finished with the paged locations you must page it out again *and* reset the interrupt. This is done with POKE 56334,129. As can be seen from the above, address one is the address which controls paging and that 55 is the normal value in that address.

We can remove the OS (operating system) by clearing bit one (mask two) of address one. This is done with POKE 1,53. However, as the computer cannot last for more than a few microseconds without it this is not a wise thing to do, but you can write your own and substitute it if you want. To switch the OS back in type POKE 1,PEEK(1) OR 2.

Flipping out the KERNAL also switches out BASIC, so bit one switches out both ROMs. BASIC can, however, be switched out on its own by clearing bit zero of address one. Flip it out and we have 8K more for our machne code programs. This is very useful if we have an all machine code program and therefore don't need BASIC.

We can copy BASIC into RAM. This lets us change BASIC if we don't like it. The following program copies BASIC into RAM behind the BASIC ROM.

```
10 FORA=40960TO49152
20 POKEA,PEEK(A)
30 NEXT
```

Line 20 may look a little stupid, but if you POKE a ROM on the 64 the value is put in the RAM directly behind the ROM. Run the program. It will take a minute or so. Now the BASIC language is in RAM, *and* ROM. Let's switch to RAM BASIC. Type POKE 1,54. If the cursor is still flashing then everything is OK. If not, turn the computer off and on then re-type and run the program. Now BASIC is in RAM, we can now adjust and change BASIC to our needs. BASIC on the Commodore 64 has a bug in it. Let's correct it.

If you type PRINT ASC ("") the 64 will give you an error message:

? ILLEGAL QUANTITY ERROR.

The value returned should be a zero. To correct the bug, type POKE 46991,5. This slightly changes the ASC function to give a zero. Everything else (including other ASC values) will work in exactly the same way.

PRINT ASC ("")

Answer . . . 0

If you want you can change BASIC commands to say what you want you can. For example, to change the BASIC command LIST to BUST, type the following:

POKE 41230,85 : POKE 41229,66

If you now type LIST you will get a syntax error, but if you type BUST the program will be listed. The BASIC commands are stored from locations 41118 to 41864 inclusive. To change any command, find the relevant address and POKE the ASC value for the letters you wish to change. The following program lets you see what character is at what address so that you know which address to POKE to change the respective command.

```
10 FORA=40960TO49152
20 POKEA,PEEK(A)
30 NEXT
```

To return to ROM BASIC at any time POKE 1,55.

In the Commodore 64 there are various interrupt vectors. These are two bypte vectors which hold the low and high byte values for subroutines which are called every sixtieth of a second by the operating system. The one that we will use is at locations 788 and 789. All of these vectors can be changed to accommodate your own machine code routine.

eg.
The vector at locations 768/769 can be changed to allow you to add new commands to BASIC. (It is the error vector so you intercept the error and do things accordingly.)

Location 788 contains the low byte and 789 the high byte.

788 normally contains 49 and 789 contains 234. You can disable the BREAK key by POKE 788,52.

The high byte is the most significant byte of the address/256. So the MSB address for this vector is 234 * 256 = 59904. As 788 contains 49, the routine which is done every sixtieth of a second is at address 49 + 59904 which is 59953.

This routine is essential for the proper running of the 64, so when your routine is finished it must always jump to 17. To change this vector for your routine change the starting address of your routine into Hex. Then take the *second* two Hex digits. Change the Hex number into decimal and POKE this number into location 788. Then take the *first* two Hex digits and change them into decimal and POKE this number into 789. However, before doing the above you need to disable the interrupts with either POKE 56334,128 in BASIC or SEI in machine code. Then, when the two numbers are POKEd into memory, type (in a program) POKE56334,129 in BASIC or CLI in machine code.

Function Keys

There are three ways of making these keys do something useful:

1. Use the CHR$ codes. This is only useful in programs where you say 'PRESS F1 to PLAY', for example. See Appendix 3 for codes.

eg.

```
10 IFA$="e"THENGOTO...:REM F1 KEY
20 IFA$=CHR$(133)THEN...:REM ALSO F1 KEY
```

2. PEEK the keyboard (locations 197 or 203). This returns a number according to which key was pressed. (For the function keys the number is between three and six.)

3. The same as above only in machine code (an interrupt routine). The routine checks the keyboard (197) to see which function key has been pressed, and then sees if the SHIFT key has been pressed. It then prints the characters associated with each key.

The following program does just that. First is a disassembly, then a BASIC program which enters the code and the characters to be printed. The ASCII codes for the characters are stored from locations 49409 to 49473 inclusive (C101 to C141 Hex).

```
C000 78             SEI
C001 A9 10          LDA #$10
C003 8D 14 03       STA $0314
C006 A9 C0          LDA #$C0
C008 8D 15 03       STA $0315
C00B 58             CLI
C00C 60             RTS
C00D EA             NOP
C00E EA             NOP
C00F EA             NOP
C010 48             PHA
C011 8A             TXA
C012 48             PHA
C013 98             TYA
C014 48             PHA
C015 A5 C5          LDA $C5
C017 C5 FB          CMP $FB
```

```
C019 F0 51       BEQ $C06C
C01B 85 FB       STA $FB
C01D C9 03       CMP #$03
C01F D0 08       BNE $C029
C021 A9 30       LDA #$30
C023 8D 00 C1    STA $C100
C026 4C 4A C0    JMP $C04A
C029 C9 04       CMP #$04
C02B D0 08       BNE $C035
C02D A9 00       LDA #$00
C02F 8D 00 C1    STA $C100
C032 4C 4A C0    JMP $C04A
C035 C9 05       CMP #$05
C037 D0 08       BNE $C041
C039 A9 10       LDA #$10
C03B 8D 00 C1    STA $C100
C03E 4C 4A C0    JMP $C04A
C041 C9 06       CMP #$06
C043 D0 27       BNE $C06C
C045 A9 20       LDA #$20
C047 8D 00 C1    STA $C100
C04A AD 8D 02    LDA $028D
C04D C9 01       CMP #$01
C04F D0 08       BNE $C059
C051 AD 00 C1    LDA $C100
C054 69 08       ADC #$08
C056 8D 00 C1    STA $C100
C059 A2 00       LDX #$00
C05B AC 00 C1    LDY $C100
C05E B9 01 C1    LDA $C101,Y
C061 9D 77 02    STA $0277,X
C064 E8          INX
C065 C8          INY
C066 E0 08       CPX #$08
C068 D0 F4       BNE $C05E
C06A 86 C6       STX $C6
C06C 68          PLA
C06D A8          TAY
C06E 68          PLA
C06F AA          TAX
```

```
C070 68            PLA
C071 4C 31 EA      JMP $EA31
```

Just type in the following BASIC program and run it. When the computer says 'READY' the function keys are programmed:

```
10 data120,169,16,141,20,3,169,192,141,2
1,3,88,96,234,234,234,72,138,72,152,72
15 data165,197,197,251,240,81,133,251,20
1,3,208,8,169,48,141,0,193,76,74,192
20 data201,4,208,8,169,0,141,0,193,76,74
,192,201,5,208,8,169,16,141,0,193,76,74
25 data192,201,6,208,39,169,32,141,0,193
,173,141,2,201,1,208,8,173,0,193,105,8
30 data141,0,193,162,0,172,0,193,185,1,1
93,157,119,2,232,200,224,8,208,244,134
35 data198,104,168,104,170,104,76,49,234
40 fora=49152to49267:readb:pokea,b:next
50 fora=0to7:readk$:forb=1to8:l=asc((mid
$(k$,b,1))):ifl=95thenl=13
55 ifl=205thenl=4
60 poke49409+(a*8)+b,l:next:next:poke494
09,4:sys49152
70 data"list-MMM"
80 data"run-MMMM"
90 data"printMMM"
100 data"thenMMMM"
110 data"loadMMMM"
120 data"saveMMMM"
130 data"verifyMM"
140 data"gotoMMMM"
```

Notes on the above program:

Lines 10–35 contain the machine code for the routine.
Line 40 POKEs in the machine code into memory. This is a 4K block of RAM which is useable in BASIC so you are still left with the full memory capacity for BASIC programs
Lines 50–60 poke the data for what will be printed when the

keys are pressed (F1 – F8). The 'M' signs are needed to fill the data up to eight characters per key, but they are changed to characters which print nothing in order to make the routine work correctly. The maximum number of characters which can be in the quotes is eight (per key).

SECTION 2.8

SIMILARITIES BETWEEN MACHINE CODE AND BASIC

Some BASIC commands can be directly changed into machine code. They are as follows:

POKE number, value	LDA value: STA number. LDX value: STX number. LDY value: STY number.
AND number/variable OR number/variable	AND number / variable. DRA number / variable.
IF variable=	CMP accumulator with. CPXX register with. CPYY register with.
THEN	BNE branch if not equal. BEQ branch if equal.
SYS address	JSR address.

SECTION 2.9

PROGRAMS TO TRY

Here is a hexadecimal loader program. It allows you to write machine code in Hex and type the Hex in. It also allows you to run the machine code, save or load it onto tape and is used to enter the Hex machine code programs in this book.

```
5 A$="0123456789ABCDEF"
10 PRINT"s"
15 POKE198,0
20 INPUT"LOAD PROGRAM";L$:IFLEFT$(L$,1)=
"Y"THENGOSUB1000:GOTO40
30 INPUT"SAVE PROGRAM";S$:IFLEFT$(L$,1)=
"Y"THENGOSUB2000
40 INPUT"STARTING LOCATION";SL
50 PRINTSL;:INPUT"HEX(2) DIGITS";H$
60 IFH$="END"THEN200
70 FORA=1TO16:B=A-1:IFLEFT$(H$,1)=MID$(A
$,A,1)THEN80
75 NEXT
77 GOTO50
80 DE=B*16
100 FORA=1TO16:B=A-1:IFRIGHT$(H$,1)=MID$
(A$,A,1)THEN120
110 NEXT:GOTO50
120 DE=DE+B
130 POKESL,DE:SL=SL+1:DE=0:H$="":GOTO50
200 INPUT"RUN MC.";R$:IFLEFT$(R$,1)="Y"T
HEN220
210 RUN
220 INPUT"STARTING LOCATION";SL:SYS(SL)
```

```
1000 INPUT"ADDRESS TO START AT";AS:INPUT
"NAME OF PROGRAM";N$
1005 INPUT"FINISHING ADDRESS";FA
1010 OPEN1,1,0,N$
1020 FORX=SATOFA:INPUT#1,Y:POKEX,Y:NEXT
1030 CLOSE1
1040 RETURN
2000 INPUT"ADDRESS TO START AT";AS:INPUT
"NAME OF PROGRAM";N$
2005 INPUT"FINISHING ADDRESS";FA
2010 OPEN1,1,2,N$
2020 FORX=SATOFA:Y=PEEK(X):PRINT#1,Y::NE
XT
2030 CLOSE1
2040 RETURN
```

Following are a Hex to decimal converter, a decimal to Hex converter and a binary to decimal converter.

```
10 INPUT"HEX";H$
20 A$="0123456789ABCDEF"
30 FORA=1TO16:B=A-1:IFLEFT$(H$,1)=MID$(A
$,A,1)THEN50
40 NEXT
50 DE=B*16
60 FORA=1TO16:B=A-1:IFRIGHT$(H$,1)=MID$(
A$,A,1)THEN80
70 NEXT
80 DE=DE+B:PRINTDE
```

```
5 A$="0123456789ABCDEF"
10 INPUT"DECIMAL NO(MAX 255)";DE
20 C=INT(DE/16):D=(DE/16-C)*16
30 H$=MID$(A$,C+1,1)+MID$(A$,D+1,1)
40 PRINTH$
50 GOTO10
```

```
10 INPUT"BINARY(8 BIT)";BN$
20 IFLEN(BN$)>8THENPRINT"TOO LONG":GOTO1
0
30 DE=0:A=0
40 FORX=8TO1STEP-1:A=A+1:IFVAL(MID$(BN$,
A,1))>1THENPRINT"ERROR":GOTO10
50 DE=DE+VAL(MID$(BN$,A,1))*2↑(X-1)
60 NEXT:PRINT"DECIMAL=";DE:GOTO10
```

Here is a maze game. You have to guide your character around the maze in the shortest possible time. Some of the lines are greater than 80 characters in length (which is the maximum allowed by the Commodore 64). so you will have to use the keyword abbreviations in order to enter the program correctly.

```
0 printchr$(142)chr$(8)
1 poke54296,15:poke54277,128:poke54278,1
28:poke54276,33
4 dima$(100):ht$=""
5 print"S□qqqqq^^^^^^^^^^^^^^^^m*a*z*e"
6 poke53281,15:poke53280,9:print"qguide
your man around the maze as fast q as po
ssible";
7 print" using keys":print"qq     rzR le
ft     rxR right":print"            qqrkR u
p rmR down"
8 print"qq^^^^^^^rpress a key to play"
9 poke198,0:wait198,1
10 fora=8448to8460:pokea,0:next
11 poke53272,24
12 print"S":poke54296,15
13 rem high res
14 poke53281,0
15 a=8712
16 readb:ifb=300then18
17 pokea,b:a=a+1:goto16
18 fora=8576to8662:readb:pokea,b:next:po
ke8199+58*8,0
20 a$(10)="AAAAAAAAAAAAAAAAAAAAAA"
```

```
21 fora=0to9:a$(a)=a$(10):next
22 a$(11)="A         eCD_         A"
23 p=1344
24 a$(12)="A AAAAAAAAAAAAAAAAAA A"
25 a$(13)="A A                A A"
26 a$(14)="A A A AA      AA A A A"
27 a$(15)="A A A  A      AA A A A"
28 a$(16)="A A A  AA    AA  A A A"
29 a$(17)="A A A   AAAAAA   A A A"
30 a$(18)="A A       AA       A A"
31 a$(19)="A AAAAA   AA   AAAAA A"
32 a$(20)="A         AA         A"
33 a$(21)="AAAAAAA A AA A AAAAA A"
34 a$(22)="A A   A A AA A A     A"
35 a$(23)="A A A A A AA A A A A A"
36 a$(24)="A A A A A AA A A A A A"
37 a$(25)="A A A A A    A A A A A"
38 a$(26)="A A A   AA  AA   A A A"
39 a$(27)="A A A    A  A    A A A"
40 a$(28)="A A AAAA A  A AAAA A A"
41 a$(29)="A      A A  A A      A"
42 a$(30)="A AAAA A      A AAAA A"
43 a$(31)="A A  A AAAAAAAA A  A A"
44 a$(32)="A A  A          A  A A"
45 a$(33)="A A  AAAAA  AAAAA  A A"
46 a$(34)="A                    A"
47 a$(35)="A AAAAAAAAAAAAAAAAAAAA"
48 a$(36)="A    A               A"
49 a$(37)="AAAA   AAAAAAAAAAAAA A"
50 a$(38)="A  AAAAA             A"
51 a$(39)="A         AA         A"
52 a$(40)="AAAAAAAA  AAAA  AAAAAA"
53 a$(41)="A     AA  AA  AA     A"
54 a$(42)="A AAA  AA    AA  AAA A"
55 a$(43)="A   A   AA  AA   A   A"
56 a$(44)="A A A            A A A"
57 a$(45)="A A AAAAAAAAAAAAAA A A"
58 a$(46)="A A                A A"
59 a$(47)="A AAAAA A    A AAAAA A"
60 a$(48)="A     A AAAAAA A     A"
```

```
61 a$(49)="AAAAA      AA      AAAAA"
62 fora=0to49:a$(50+a)=a$(49-a)
63 next
65 a$(88)="A                        A"
66 goto72
67 data255,129,255,129,255,129,255,129
68 data129,90,36,126,255,60,36,195,255,1
27,63,31,15,7,3,1,128,192,224,240,248
69 data252,254,255,0,126,24,24,24,24,24,
24,0,60,24,24,24,24,24,60
70 data0,126,64,64,126,64,64,126,0,66,10
2,90,66,66,66,66,0
71 data300
72 a=86:x=20:ti$="000000"
73 print"sqqqqq"
74 forb=atoa+4
75 printtab(8);a$(b)
76 next
77 ifpeek(p+x)<>32then88
78 pokep+54272+x,5:pokep+x,66
79 print"seqqqqqqqqqqqqqqq^^^^^^^^^^^^^^^E
FHG:";mid$(ti$,3,2)":";right$(ti$,2);"_
80 e=peek(197):ife=64then73
81 poke54272,100:poke54273,100
82 ife=36thena=a+1:m=1
83 ife=37thena=a-1:m=2
84 poke54272,0:poke54273,0
85 ife=12thenx=x-1:m=3
86 ife=23thenx=x+1:m=4
87 goto73
88 ifm=1thena=a-1
89 ifpeek(p+x)=67orpeek(p+x)=68then96
90 poke54272,100:poke54273,100
91 ifm=2thena=a+1
92 ifm=3thenx=x+1
93 ifm=4thenx=x-1
94 poke54272,0:poke54273,0
95 goto73
96 fora=255to0step-4:forb=5to15step3
97 poke54296,b:poke54272,200:poke54273,a
```

```
99 nextb,a:poke54272,0:poke54273,0:poke5
3272,21
100 print"Serqqqqq^^your time was ";mid$
(ti$,3,2);" minutes ";right$(ti$,2);" se
conds"
101 print"sqqqqqqqqqqqqqqqqqqqq^^^^^^^^^^
reany key to restart_"
102 ifpeek(197)<>64then102
103 ifpeek(197)=64then103:print"S"
104 ti$="000000"
106 restore
107 print"sqqqqqqqqqqqqq"
108 run
110 data0,126,70,74,82,98,126,0,0,56,24,
24,24,24,126,0,0,126,66,4,120,98,126,0
112 data0,126,2,62,6,6,126,0,0,64,64,64,
96,100,126,4,0,126,64,64,126,6,126,0
114 data0,96,96,126,98,98,126,0,0,126,2,
2,6,6,6,0,0,126,66,126,66,66,126,0
116 data0,126,66,66,126,6,6,0,0,0,24,0,0
,24,0,0,0
118 data0,126,24,24,24,24,24,24,0
```

Here is an adventure program that takes place in the underground caves of Marple.

```
1 printchr$(14)chr$(8):poke53280,2:poke5
3281,0
10 print"S↑":dima(425)
20 print:print
30 print"Sq              Treasure Hunt.":f
orhh=1to2500:next
40 print:print
60 deffnr(x)=int(rnd(1)*x)+1
70 print"SqqThe MARPLE CAVES,located in
Marple-town,are said to have 20"
71 print"Hidden treasures in them.Few ex
plore    them,because it is said that"
```

```
72 print"PIRATES and DRAGONS live there-
-besides"
80 print"there are deep pits which many
have      fallen into and DIED"
81 print"qqPRESS f1 TO CONTINUE"
83 getm$
85 ifm$="E"then88
86 goto83
88 print"Sqqqq"
90 print"YOU,a smart and brave human,and
 I a      Commodore 64 will explore the c
aves"
91 print"and try to find the   treasure.
I will be your eyes and ears"
92 print"and will tell you if danger lur
ks ahead."
100 print
110 print"  I hope you brought a map.In
case you  didn't,you'll have to"
111 print"make one as we explore":print:
print"qqPRESS f1 TO CONTINUE."
112 getm$
113 ifm$="E"then140
115 goto112
140 print"S"
141 fori=1to3
150 n=3+fnr(88)
160 ifa(n)<>0then150
170 a(n)=2
180 next
190 fori=3to9
200 n=3+fnr(88)
210 ifa(n)<>0then200
220 a(n)=i
230 next
240 fori=288to304
250 n=fnr(94)
260 ifa(n)<>0then250
270 a(n)=i
280 next
```

```
290 restore
300 forn=1to23
310 reada$
320 next
330 forn=137to408
340 reada(n)
350 next
360 w=0:m=0:b=200:d=0
400 print"qqqq"
410 print" Welcome to the Marple Caves w
here you will find MYSTERY"
411 print"TREASURE and ADVENTURE!"
420 print
430 goto500
470 print"S"
480 b=b-1
490 ifb=0then3740
500 ifw=0then3140
540 iffnr(200)=200thengosub1890
550 iffnr(200)=200thengosub2120
560 ifd=0anda(w)=0andfnr(50)=50thengosub
2850
570 ifa(w)=6and(fnr(5)=5)then3570
620 ifa(w)=2then3440
630 ifa(w)=4then2890
640 ifa(w)=5thengosub1770
650 ifa(w)=7thenprint:print:print"There'
s a shy little  elf in here hiding    so
mething.
660 ifa(w)=8then2410
700 gosub3350
710 z=w
720 gosub4120
730 forn=134to136
750 printtab(1)"CAVE";a(n);
760 ifa(n)=0thenprint"-THE ENTRANCE!";
770 print
780 next
820 ifb>190then880
830 ifw<>0then880
```

```
840 input"To explore hit1 else 2"
860 ifj=2then3810
870 ifj<>1then840
880 ifb=50thenprint:print"  I hope you b
rought  your torch batteries"
890 ifb=25thenprint:print"Your torch is
         starting to dim a bit."
900 ifb=8thenprint:print"Your torch is a
lmost  out. Better leave     right now!!
!!
910 ifa(w)=3thengos2290
950 k=0:l=0
970 forn=134to136
980 j=a(w)
990 ifj<0then1090
1000 ifa(j)=5then1090
1010 ifa(j)>0anda(j)<10andk=0thenprint:k
=1
1020 ifa(j)=2andl=0thenprint"▲▲There are
 pits       nearby.watch your step.";l=1
1030 ifa(j)=3thenprint" CAREFUL- There's
 a  pirate nearby"
1040 ifa(j)=4thenprint" I hear a hungry
     dragon nearby-waiting  for his SUPP
ER!"
1050 if(j)=6thenprint"There's a sign her
e   that says:"print"Sggg▲▲r >>DANGER<<R
□"
1060 ifa(j)=7thenprint"Sounds like someo
ne is singing. Must be an  ELF."
1070 ifa(j)=8thenprint"There's a strange
 mist in here!"
1080 ifa(j)=9thenprint"RUFF RUFF!.... I
can  hear a dog barking"
1090 next
1100 ifa(131)=0then1200
1110 print
1120 print"You're carrying:";a$
1140 ifa(n)<100then1190
1150 z=a(n)
```

```
1160 gosub4040
1170 printa$;
1190 print
1200 ifa(w)<100thenprint:goto1440
1240 print
1250 z=a(w)
1260 gosub4040
1270 print"▲This cave has";a$;"in it"
1280 print"Do you want to take it with y
ou?q"
1290 input"(Y orN)";j$
1300 ifj$="n"then1440
1310 ifj$<>"y"then1290
1320 ifz=290thengosub1630
1330 ifz=290then1440
1340 ifz=291thenm=1
1350 ifz=294thengosub1990
1360 ifz=1then1440
1370 ifz=295then3650
1380 forn=131to133
1390 ifa(n)=0thena(n)=a(w):a(w)=0:goto14
30
1400 next
1410 print"▲You can't pick it up now...y
ou're carrying too much."
1420 goto1440
1430 print"O.K.you've got";a$"."
1440 ifa(w)=7thena(w)=307
1480 input"which cave next";h
1490 forn=134to136
1500 ifa(n)=-1then1520
1510 ifh=a(n)thenw=h:goto470
1520 next
1530 ifm=1andh>=0andh<=94andint(h)=hthen
w=h:goto470
1540 print"q Sorry,but you can't  go the
re from here.q"
1550 goto1480
1630 forn=131to133
1640 ifa(n)=298then1680
```

```
1650 next
1660 print"You can't take it-it's too he
avy!"
1670 return
1680 print" The box was too heavy so I o
pened it with your keys.By the way,there
 is a"
1681 print"_ruby□ in it that you can tak
e."
1690 a(w)=308
1700 z=308
1710 gosub4040
1720 return
1760 print"qqqq"
1770 print" There's a FRUIT      MACHINE
 in here."
1780 forn=131to133
1790 ifa(n)=288then1820
1800 next
1810 return
1820 print"I'll get some         batteri
es with one of your ↑GOLD□ coins."
1830 b=200
1840 a(w)=0
1850 return
1890 forn=1to10
1900 z=3+fnr(88)
1910 gosub3950
1920 next
1930 print"qqqqqq"
1940 print"Guess what! We've     just had
 an earthquake!!But I'm O.K..."
1950 return
1990 forn=131to133
2000 ifa(n)=293then2040
2010 next
2020 print" It's stuck in the    rocks a
nd can't be    pulled out."
2030 z=1:return
2040 print" Let me use your      magic w
```

```
and a sec.";
2050 forz=0to2000:next
2060 print"qq HOCUS,POCUS..."
2070 forz=0to2000:next
2080 a(w)=310:z=0
2090 return
2120 print"qqqqqq":print""
2130 print" A superbat just flew cave an
d picked you up."
2140 ifa(131)<>0thenprint"You dropped al
l your  treasures"
2150 forn=131to133
2160 ifa(n)=0then2180
2170 a(0)=a(n):a(n)=0:z=0:gosub3950
2180 next
2190 n=fnr(94)
2200 ifa(n)<>0then2190
2210 w=n
2220 print" The bat just dropped you int
o cave";w"!!!"
2230 print"If you're all right hit a key
."
2231 geta$:ifa$=""then2231
2233 print"S"
2240 print"s"
2250 return
2290 print"qq There was a pirate in here
."
2300 ifa(13)<>0thenprint"HE JUST STOLE A
LL YOUR TREASURES!"
2310 forn=131to133
2320 a(w)=0
2330 next
2340 z=w
2350 gosub3950
2360 print" HE'S GONE NOW!."
2370 return
2400 print"S"
2410 print"qqqqqq":print
2420 print"There's a magician in this ca
```

```
ve.He says he's lost his magic book."
2421 print"He says he'll give you a ↑GOLDEN HARP□ if you'll tell him where it is."
2422 print"qq▲Type in the cave  number.If you don't know then just type 99.q"
2450 input"Where's his book?";j
2470 ifj<0andj>95then2520
2480 ifa(j)<>295then3840
2490 a(w)=309
2500 a(j)=0
2510 goto470
2520 z=w
2530 gosub3950
2540 goto470
2580 print"qqqqqq";print
2590 print"The invisible man is  here looking for his  invisible dog.He says he'll give
2595 print"you a $1000 reward if you can tell him where it is.If you don't▲▲know"
2596 "Q▲▲▲▲▲then guess.q"
2610 input"What cave is it inq";j$
2620 print"S"
2630 j=int(val(j$))
2640 ifj<0andj>95andint(j)<>janda(j)<>9thenreturn
2650 a(w)=305
2660 a(j)=0
2670 d=1
2680 return
2720 forn=131to133
2730 ifa(n)=291then2770
2740 next
2750 m=0
2760 return
2770 a(w)=0
2780 ifn=133then2830
2790 forj=nto132
```

```
2800 a(j)=a(j+1)
2810 a(j+1)=0
2820 next
2830 print" PUFF!!! Your magic   carpet
just diappeared."
2840 m=0:return
2850 rem
2890 forn=131to133
2900 ifa(n)=292then2930
2910 next
2920 goto3500
2930 print"qqqqqq":print"YIKES-there's a
 dragon in here."
2950 print"Give me your gun'qAAAA QUICK!!
!"
2960 forn=0to2000:next
2970 print"S"
2980 forn=1to5
2990 printtab(5)"qqqBANG!!!"
3000 forj=0to500:next:print"S"
3010 next
3020 printtab(fnr(30));"PUFF!!"
3030 forj=0to300:next:print"S"
3040 print"INCREDIBLE ???The     dragon
just vanished  when I shot him right bet
ween the"
3041 print"q eyes□"
3050 print:print" But look at this-he  l
eft his little black book behind-with th
e "
3051 print"EVERY BEAUTIFUL        PRINCES
S IN PENNSLY-]]VANIA."
3052 print"address of every beautiful pr
incess in PENNYSLYVANIA!"
3070 print"qHit a key when you  catch yo
ur breath"
3071 geta$:ifa$=""then3071
3080 a(w)=306
3090 b=b+1
3100 goto470
```

```
3140 ifa(131)=0then3210
3150 forn=131to133
3160 ifa(n)=0then3200
3170 j=96
3180 ifa(j)<>0thenj=j+1:goto3180
3190 a(j)=a(w)
3200 next
3210 ifa(96)=0then3270
3220 print"So far you've found these tre
asures in the caves:"
3230 forn=96to130
3240 ifa(n)=0then3270
3250 z=a(n):gosub4040:printa$",";
3260 next
3270 forn=131to133:a(n)=0:next
3280 print"qq"
3290 print"qqYou're at a cave    entranc
e that leads to:"
3300 b=b+1
3310 goto710
3350 print"qqqYou're in cave"w"which lea
ds to:"
3360 return
3440 print"qqqqqq△Sorry,  but I tried to
 warn you."
3450 print"You fell into a deep  pit ...
and _KILLED□  yourself!!!"
3460 end
3500 print"qqqqqq":print""
3510 print" Sorry but I tried to tell yo
u about that sound.Supper is now being s
erved he
3511 print"in the"
3512 print"πDRAGON'S CHAMBER..."
3520 print"q":fori=1to2500:next:print"qq
rAND YOU ARE THE qq△   SUPPER!!!□"
3530 end
3570 print"qqqqqq":print""
3580 print"The roof just fell in and bur
ied you alive. Too bad, I don't have a s
```

```
hovel"
3581 print"or I'd dig you out."
3600 print"qSee you later."
3610 end
3650 print"sqqqqqqq"
3670 print"I don't think you .   should
have done that!The magician who owns tha
t "
3672 print"Q^^^^^magic book"
3675 print"put a spell on it.Anybody tha
t tries to pick it up turns into a πfrog
□."
3680 print
3690 print"Excuse me while I look for so
me flies."
3700 end
3740 print"qqqqqq":print""
3750 print" Now look what you did.Your f
lashlight went out and you fell into a p
it and"
3751 print"KILLED YOURSELF"
3755 print"TOO BAD! Especially as you we
re doing so well."
3760 end
3800 print"qqqqqqThe treasures are yours
 to keep!WELL DONE!!!"
3810 end
3840 print"qqqqqq You've made the magici
an VERY angry.. Cave ";j;"doesn't have a
 "

3850 print"magic book in it."
3860 print"To punish you,the     magicia
n casts a spell on you and now you're on
ly"
3861 print"TWO INCHES TALL!!!"
3870 print"Worse yet,the magician put yo
u in a small   jar.If you ever get  out
of this"
3871 print"messqllllLET ME KNOW!!!"
3880 end
```

```
3950 y=3+fnr(88)
3960 ifa(y)<>0then3950
3970 ify=wandy=zthen3950
3980 a(y)=a(z)
3990 a(z)=0
4000 return
4040 restore
4050 forx=0toz-288
4060 reada$
4070 next
4080 return
4120 forx=0to2
4130 a(134+x)=a(137+x+(z*3))
4140 next
4150 return
9910 data" gold coins "
9911 data" some keys "
9912 data" a locked box "
9913 data" a magic carpet "
9914 data" an old gun "
9915 data" a magic wand "
9916 data" a sword "
9917 data" a magic book "
9918 data" an old clock "
9919 data" some furs "
9920 data" a silver bell "
9921 data" a         necklace  "
9922 data".a pearl "
9923 data" a diamond "
9924 data" a gold watch "
9925 data" an emerald "
9926 data" some jewellry "
9927 data" a _1000 note "
9928 data" a black book "
9929 data" some elf food "
9930 data" a ruby "
9931 data" a golden harp "
9932 data" a sword "
10000 data1,94,1,0,2,3,1,4;5,1,6,7,2,8,9
.2,10,11,3,12,13,3,14,15,4,16,17,4,18,19
```

```
10001 data5,20,21,5,22,23,6,24,25,6,26,2
7,7,28,29,7,30,31,7,32,33,8,34,35,9,36,3
7
10011 data9,38,39,10,40,41,10,42,43,11,4
4,45,11,44,45,11,46,47,12,48,49,12,50,51
10023 data13,52,53,13,54,55,14,56,57,14,
58,59,15,60,61,15,62,63,16,63,64,16,34,6
4
10024 data17,33,65,17,36,37,18,35,66,18,
38,66,19,37,67,19,40,67,20,39,68,20,42,6
8
10025 data21,41,69,21,44,69,22,43,70,22,
46,70,23,45,71,23,48,71,24,47,72,24,50,7
2
10026 data25,49,73,25,52,73,26,51,74,26,
54,78,74,27,53,75,27,56,75,28,55,76,28,5
8
10027 data76
10028 data29,57,77,29,60,77,30,59,78,30,
62,78,31,61,79,31,32,79,32,33,80,34,35,8
,0
10029 data36,37,81,38,39,81,40,41,82,42,
43,82,44,45,83,46,47,83,48,49,84,50,51,8
,4
10030 data52,53,85,54,55,85,56,57,86,58,
59,86,60,61,87,62,63,87,64,65,88,66,67,8
8
10031 data68,69,89,70,71,89,72,73,90,74,
75,90,76,77,91,78,79,91,80,81,92,82,83,9
2
```

APPENDICES

APPENDIX 1

Abbreviations for BASIC keywords

Commodore 64 BASIC allows you to abbreviate most of the BASIC keywords. The usual format for these are the first letter of the keyword and the second leter shifted. When you LIST the program the keywords will be listed out in full.

Command	Abbreviation	Looks like this on screen
ABS	A SHIFT B	A
AND	A SHIFT	A ╱
ASC	A SHIFT S	A ♥
ATN	A SHIFT T	A
CHR$	C SHIFT H	C
CLOSE	CL SHIFT O	CL
CLR	C SHIFT L	C
CMD	C SHIFT M	C ╲
CONT	C SHIFT O	C
DATA	D SHIFT A	D ♠
DEF	D SHIFT E	D
DIM	D SHIFT I	D

END	E SHIFT N	E
EXP	E SHIFT X	E ♣
FOR	F SHIFT O	F
FRE	F SHIFT R	F
GET	G SHIFT E	G
GOSUB	GO SHIFT S	GO ♥
GOTO	SHIFT O	G
INPUT#	I SHIFT N	I
LET	L SHIFT E	L
LEFT$	LE SHIFT F	LE
LIST	L SHIFT I	L
LOAD	L SHIFT O	L
MID$	M SHIFT I	M
NEXT	N SHIFT E	N
NOT	N SHIFT O	N
OPEN	O SHIFT P	O
PEEK	P SHIFT E	P
POKE	P SHIFT O	P
PRINT	?	?
PRINT#	P SHIFT R	P
READ	R SHIFT E	R
RESTORE	RE SHIFT S	RE ♥
RETURN	RE SHIFT T	RE
RIGHT$	R SHIFT I	R
RND	R SHIFT N	R

RUN	R	SHIFT		R
SAVE	S	SHIFT	A	S
SGN	S	SHIFT	G	S
SIN	S	SHIFT	I	S
SPC(	S	SHIFT	P	S
SQR	S	SHIFT	Q	S
STEP	ST	SHIFT	E	ST
STOP	S	SHIFT	T	S
STR$	ST	SHIFT	R	ST
SYS	S	SHIFT	Y	S
TAB	T	SHIFT	A	T
THEN	T	SHIFT	H	T
USR	U	SHIFT	S	U
VAL	V	SHIFT	A	V
VERIFY	V	SHIFT	E	V
WAIT	W	SHIFT	A	W

APPENDIX 2

Screen display codes

The following table lists all the characters which can be displayed on the screen along with their respective number to POKE. For example, POKE 1Ø24,1 puts an A in the top left corner of the screen. There are two character sets available to you. They cannot be used together but can be switched by hand by pressing the SHIFT and Commodore keys together, or typing POKE 53272,21 to switch to upper case (set 1) mode and POKE 53272,22 switches to lower case mode.

SET 1	SET 2	POKE
@		0
A	a	1
B	b	2
C	c	3
D	d	4
E	e	5
F	f	6
G	g	7
H	h	8
I	i	9
J	j	10
K	k	11
L	l	12
M	m	13
N	n	14
O	o	15
P	p	16
Q	q	17
R	r	18
S	s	19
T	t	20
U	u	21
V	v	22
W	w	23
X	x	24
Y	y	25
Z	z	26

[		27
£		28
]		29
↑		30
←		31
SPACE		32
!		33
"		34
#		35
$		36
%		37
&		38
'		39
(		40
)		41
*		42
+		43
,		44
–		45
.		46
/		47
0		48
1		49
2		50
3		51
4		52
5		53
6		54
7		55
8		56
9		57
:		58
;		59
<		60
=		61
>		62
?		63
		64
	A	65
	B	66
	C	67
	D	68
	E	69
	F	70
	G	71
	H	72
	I	73
	J	74
	K	75
	L	76
	M	77
	N	78
	O	79
	P	80
	Q	81
	R	82

S	83
T	84
U	85
V	86
W	87
X	88
Y	89
Z	90
	91
	92
	93
	94
	95
	96
	97
	98
	99
	100
	101
	102
	103
	104

105
106
107
108
109
110
111
112
113
114
115
116
117
118
119
120
121
122
123
124
125
126
127

Codes 128–255 are reversed images of code Ø–127.

APPENDIX 3

ASCII and CHR$ codes

This table shows you what character will be displayed if you type PRINT CHR$ (X). It also has the values you will obtain if you type PRINT ASC ("X") where X is any character you can type. The CHR$ codes also change case (lower to upper, etc), change colours control the printer, etc...

PRINTS	CHR$	PRINTS	CHR$	PRINTS	CHR$	PRINTS	CHR$
	0	CRSR ↓	17	"	34	3	51
	1	RVS ON	18	#	35	4	52
	2	CLR HOME	19	$	36	5	53
	3	INST DEL	20	%	37	6	54
	4		21	&	38	7	55
WHT	5		22	'	39	8	56
	6		23	(	40	9	57
	7		24	)	41	:	58
DISABLES SHIFT C=	8		25	*	42	;	59
ENABLES SHIFT C=	9		26	+	43	<	60
	10		27	,	44	=	61
	11	RED	28	–	45	>	62
	12	CRSR →	29	.	46	?	63
RETURN	13	GRN	30	/	47	@	64
SWITCH TO LOWER CASE	14	BLU	31	0	48	A	65
	15	SPACE	32	1	49	B	66
	16	!	33	2	50	C	67

PRINTS	CHR$	PRINTS	CHR$	PRINTS	CHR$	PRINTS	CHR$
D	68	♠	97		126	GREY 3	155
E	69		98		127	PUR	156
F	70		99		128	CRSR	157
G	71		100	ORANGE	129	YEL	158
H	72		101		130	CYN	159
I	73		102		131	SPACE	160
J	74		103		132		161
K	75		104	f1	133		162
L	76		105	f3	134		163
M	77		106	f5	135		164
N	78		107	f7	136		165
O	79		108	f2	137		166
P	80		109	f4	138		167
Q	81		110	f6	139		168
R	82		111	f8	140		169
S	83		112	SHIFT RETURN	141		170
T	84		113	SWITCH TO UPPER CASE	142		171
U	85		114		143		172
V	86	♥	115	BLK	144		173
W	87		116	CRSR	145		174
X	88		117	RVS OFF	146		175
Y	89		118	CLR HOME	147		176
Z	90		119	INST DEL	148		177
[	91	♣	120	BROWN	149		178
£	92		121	LT. RED	150		179
]	93	♦	122	GREY 1	151		180
↑	94		123	GREY 2	152		181

←	95	◧	124	LT GRN.	153	◨	182
⊟	96	◫	125	LT. BLU	154	▔	183

⬒	184	□	186	▝	188	▘	190
⬓	185	▖	187	◰	189	▚	191

CODES	**192-223**	**SAME AS**	**96-127**
CODES	**224-254**	**SAME AS**	**160-190**
CODE	**255**	**SAME AS**	**126**

APPENDIX 4

Screen and colour memory maps

The following charts show what number to POKE to put a character on the screen and colour it in any of the 16 available colours.

SCREEN MEMORY MAP

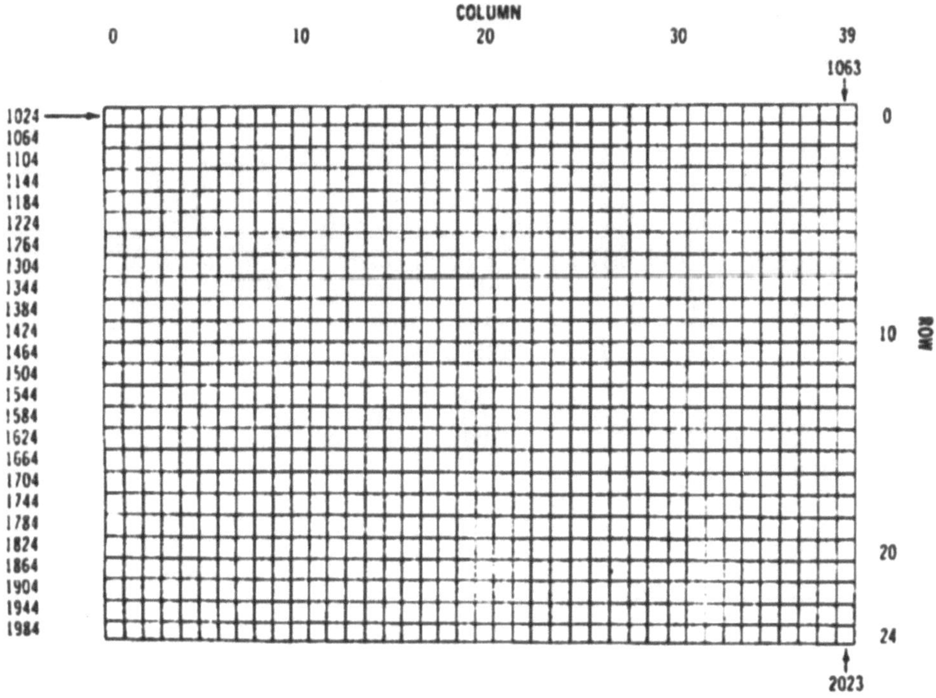

COLOR MEMORY MAP

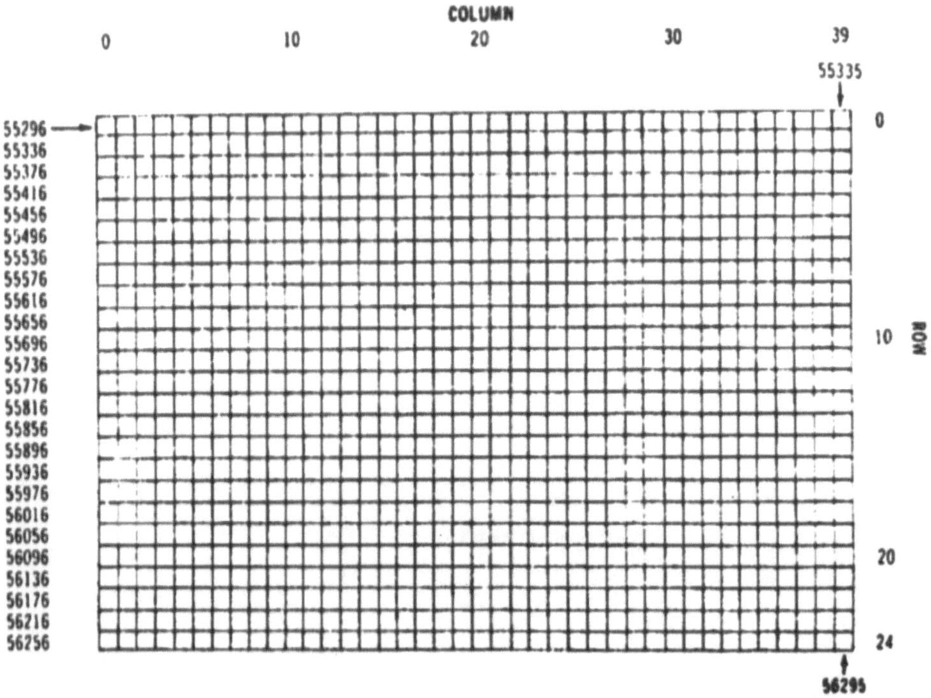

The values to POKE into the above colour map are as follows:

BLACK Ø	BLUE 6	Light RED 1Ø	Light BLUE 14
WHITE 1	YELLOW 7	GRAY 1 11	GRAY 3 15
RED 3	ORANGE 8	GRAY 2 12	
PURPLE 4	BROWN 9	Light GREEN 13	
GREEN 5			

APPENDIX 5

Deriving mathematical functions

Functions which are not standard in Commodore 64 BASIC can be calculated as follows:

FUNCTION	BASIC EQUIVALENT
SECANT	SEC(X)= 1/COS(X)
COSECANT	CSC(X)= 1/SIN(X)
COTANGENT	COT(X)= 1/TAN(X)
INVERSE SINE	ARCSIN(X)=ATN(X/SQR(−X*X+1))
INVERSE COSINE	ARCCOS(X)= −ATN(X/SQR (−X*X +1)) +π/2
INVERSE SECANT	ARCSEC(X)=ATN(X/SQR(X*X−1))
INVERSE COSECANT	ARCCSC(X)=ATN(X/SQR(X*X−1)) +(SGN(X)−1*π/2
INVERSE COTANGENT	ARCOT(X)=ATN(X)+π/2
HYPERBOLIC SINE	SINH(X)=(EXP(X)−EXP(−X))/2
HYPERBOLIC COSINE	COSH(X)=(EXP(X)+EXP(−X))/2
HYPERBOLIC TANGENT	TANH(X)=EXP(−X)/(EXP(x)+EXP (−X))*2+1
HYPERBOLIC SECANT	SECH(X)=2/(EXP(X)+EXP(−X))
HYPERBOLIC COSECANT	CSCH(X)=2/(EXP(X)−EXP(−X))
HYPERBOLIC COTANGENT	COTH(X)=EXP(−X)/(EXP(X) −EXP(−X))*2+1
INVERSE HYPERBOLIC SINE	ARCSINH(X)=LOG(X+SQR(X*X+1))
INVERSE HYPERBOLIC COSINE	ARCCOSH(X)=LOG(X+SQR(X*X−1))
INVERSE HYPERBOLIC TANGENT	ARCTANH(X)=LOG((1+X)/(1−X))/2
INVERSE HYPERBOLIC SECANT	ARCSECH(X)=LOG((SQR (−X*X+1)+1/X)
INVERSE HYPERBOLIC COSECANT	ARCCSCH(X)=LOG((SGN(X)*SQR (X*X+1/x)
INVERSE HYPERBOLIC COTAN-GENT	ARCCOTH(X)=LOG((X+1)/(x−1))/2

APPENDIX 6

Pinouts for Input/Output devices

The following charts show what can be connected to your Commodore 64 and where.

1) Game I/O
2) Cartridge Slot
3) Audio/Video
4) Serial I/O (Disk/Printer)
5) Modulator Output
6) Cassette
7) User Port

Control Port 1

Pin	Type	Note
1	JOYA0	
2	JOYA1	
3	JOYA2	
4	JOYA3	
5	POT AY	
6	BUTTON A/LP	
7	+5V	MAX. 50mA
8	GND	
9	POT AX	

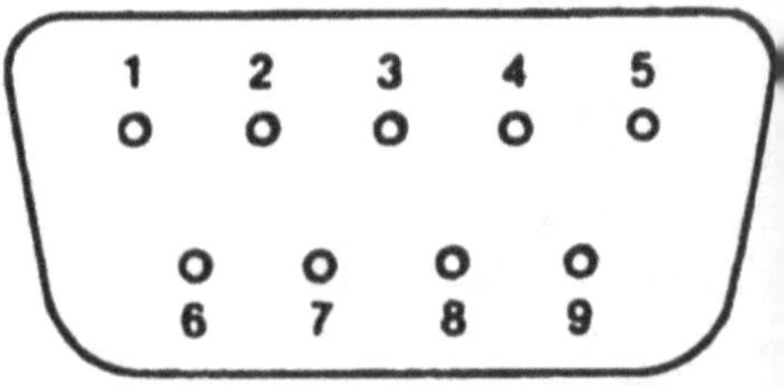

Control Port 2

Pin	Type	Note
1	JOYB0	
2	-JOYB1	
3	JOYB2	
4	JOYB3	
5	POT BY	
6	BUTTON B	
7	+5V	MAX. 50mA
8	GND	
9	POT BX	

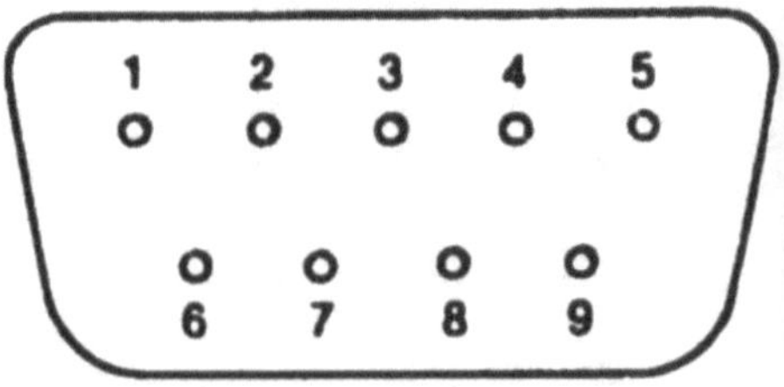

Cartridge Expansion Slot

Pin	Type
22	GND
21	CD0
20	CD1
19	CD2
18	CD3
17	CD4
16	CD5
15	CD6
14	CD7
13	DMA
12	BA

Pin	Type
11	ROML
10	I/02
9	EXROM
8	GAME
7	I/01
6	Dot Clock
5	CR/W
4	IRQ
3	+5V
2	+5V
1	GND

Pin	Type
Z	GND
Y	CA0
X	CA1
W	CA2
V	CA3
U	CA4
T	CA5
S	CA6
R	CA7
P	CA8
N	CA9

Pin	Type
M	CA10
L	CA11
K	CA12
J	CA13
H	CA14
F	CA15
E	S02
D	NMI
C	RESET
B	ROMH
A	GND

Audio/Video

Pin	Type	Note
1	LUMINANCE	
2	GND	
3	AUDIO OUT	
4	VIDEO OUT	
5	AUDIO IN	

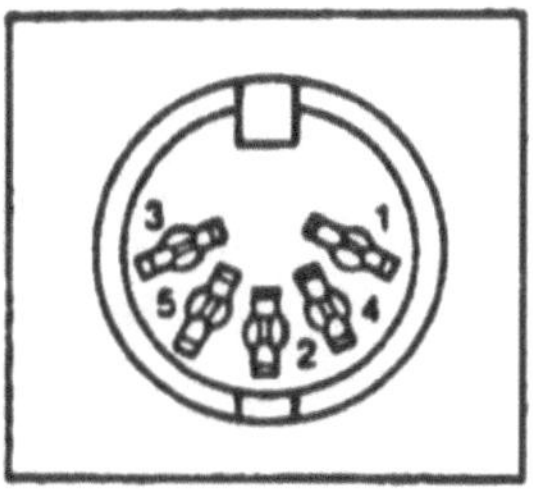

Serial I/O

Pin	Type
1	SERIAL $\overline{\text{SRQIN}}$
2	GND
3	SERIAL ATN IN/OUT
4	SERIAL CLK IN/OUT
5	SERIAL DATA IN/OUT
6	$\overline{\text{RESET}}$

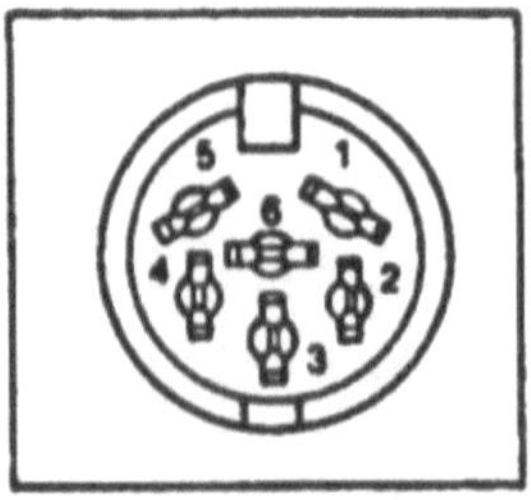

Cassette

Pin	Type
A-1	GND
B-2	+5V
C-3	CASSETTE MOTOR
D-4	CASSETTE READ
E-5	CASSETTE WRITE
F-6	CASSETTE SENSE

User I/O

Pin	Type	Note
1	GND	
2	+5V	MAX. 100 mA
3	$\overline{\text{RESET}}$	
4	CNT1	
5	SP1	
6	CNT2	
7	SP2	
8	$\overline{\text{PC2}}$	
9	SER. ATN IN	
10	9 VAC	MAX. 100 mA
11	9 VAC	MAX. 100 mA
12	GND	

Pin	Type	Note
A	GND	
B	$\overline{\text{FLAG2}}$	
C	PB0	
D	PB1	
E	PB2	
F	PB3	
H	PB4	
J	PB5	
K	PB6	
L	PB7	
M	PA2	
N	GND	

APPENDIX 7

Error message

This appendix contains a list of all the error messages generated by the Commodore 64 and the reasons that they occur.

BAD DATA String data was received from an open file, but the program was expecting numeric data.

BAD SUBSCRIPT The program was trying to reference an element of an array whose number is outside of the range specified in the DIM statement.

CAN'T CONTINUE The CONT command will not work, either because the program was never RUN, there has been an error or a line has been edited.

DEVICE NOT PRESENT The required I/O device was not available for an OPEN, CLOSE, CMD, PRINT #, INPUT # or GET #.

DIVISION BY ZERO Division by zero is a mathematical oddity and not allowed.

EXTRA IGNORED Too many items of data were typed in response to an INPUT statement. Only the first few items were accepted.

FILE NOT FOUND If you were looking for a file on tape and an 'END-OF-TAPE' marker was found. If you were looking on disc, and no file with that name exists.

FILE NOT OPEN The file specified in a CLOSE, CMD, PRINT #, INPUT # or GET #, must first be OPENed.

FILE OPEN An attempt was made to open a file using the number of an already open file.

FORMULA TOO COMPLEX The string expression being evaluated should be split into at least two parts for the system to work with or a formula has too many parenthesises.

ILLEGAL DIRECT The INPUT statement can only be used within a program, and not in direct mode.

ILLEGAL QUANTITY A number used as the argument of a function or statement is out of the allowable range.

LOAD There is a problem with the program on tape.

NEXT WITHOUT FOR This is caused by either incorrectly nesting loops or having a variable name in a NEXT statement which doesn't correspond with one in a FOR statement.

NOT INPUT FILE An attempt was made to INPUT or GET data from a file which was specified to be for output only.

NOT OUTPUT FILE An attempt was made to PRINT data to a file which was specified as input only.

OUT OF DATA A READ statement was executed but there is no data left unREAD in a DATA statement.

OUT OF MEMORY There is no more RAM available for program or variables. This may also occur when too many FOR loops have been nested, or when there are too many GOSUBs in effect.

OVERFLOW The result of a computation is larger than the largest number allowed, which is 1.7Ø141884E+38.

REDIM'D ARRAY An array may only be DIMensioned once. If an array variable is used before that array is DIM'd, an automatic DIM operation is performed on that array setting the number of elements to ten, and any subsequent DIMs will cause this error.

REDO FROM START Character data was typed in during an INPUT statement when numeric data was expected. Just re-type the entry so that it is correct, and the program will continue by itself.

RETURN WITHOUT GOSUB A RETURN statement was encountered, and no GOSUB command has been issued.

STRING TOO LONG A string can contain up to 255 characters.

?SYNTAX ERROR A statement is unrecognizable by the Commodore 64. A missing or extra parenthesis, misspelled keywords, etc.

TYPE MISMATCH This error occurs when a number is used in place of a string, or *vice versa*.

UNDEF'D FUNCTION A user defined function was referenced, but it has never been defined using the DEF FN statement.

UNDEF'D STATEMENT An attempt was made to GOTO or GOSUB or RUN a line number which doesn't exist.

VERIFY The program on tape or disc does not match the program currently in memory.

APPENDIX 8

Music note values

The following table contains a list of all the notes, their sharps and octaves and the respective numbers to POKE in order to obtain the notes. (POKE into the low and high frequency registers of the sound chip.)

Note	Note–Octave	Hi Freq	Low Freq
0	C–0	1	18
1	C#–0	1	35
2	D–0	1	52
3	D#–0	1	70
4	E–0	1	90
5	F–0	1	110
6	F#–0	1	132
7	G–0	1	155
8	G#–0	1	179
9	A–0	1	205
10	A#–0	1	233
11	B–0	2	6
12	C–1	2	37
13	C#–1	2	69
14	D–1	2	104
15	D#–1	2	140
16	E–1	2	179
17	F–1	2	220
18	F#–1	3	8
19	G–1	3	54
20	G#–1	3	103
21	A–1	3	155
22	A#–1	3	210
23	B–1	4	12
24	C–2	4	73
25	C#–2	4	139
26	D–2	4	208

27	D#–2	5	25
28	E–2	5	103
29	F–2	5	185
30	F#–2	6	16
31	G–2	6	108
32	G#–2	6	206
33	A–2	7	53
34	A#–2	7	163
35	B–2	8	23
36	C–3	8	147
37	C#–3	9	21
38	D–3	9	159
39	D#–3	10	60
40	E–3	10	205
41	F–3	11	114
42	F#–3	12	32
43	G–3	12	216
44	G#–3	13	156
45	A–3	14	107
46	A#–3	15	70
47	B–3	16	47
48	C–4	17	37
49	C#–4	18	42
50	D–4	19	63
51	D#–4	20	100
52	E–4	21	154
53	F–4	22	227
54	F#–4	24	63
55	G–4	25	177
56	G#–4	27	56
57	A–4	28	214
58	A#–4	30	141
59	B–4	32	94
60	C–5	34	75
61	C#–5	36	85
62	D–5	38	126
63	D#–5	40	200
64	E–5	43	52
65	F–5	45	198
66	F#–5	48	127
67	G–5	51	97
68	G#–5	54	111
69	A–5	57	172
70	A#–5	61	126
71	B–5	64	188

72	C-6	68	149
73	C#-6	72	169
74	D-6	76	252
75	D#-6	81	161
76	E-6	86	105
77	F6	91	140
78	F#-6	96	254
79	G-6	102	194
80	G#-6	108	223
81	A-6	115	88
82	A#-6	122	52
83	B-6	129	120
84	C-7	137	43
85	C#-7	145	83
86	D-7	153	247
87	D#-7	163	31
88	E-7	172	210
89	F-7	183	25
90	F#-7	193	252
91	G-7	205	133
92	G#-7	217	189
93	A-7	230	176
94	A#-7	244	103

MEMORY MAP

HEX	DECIMAL	DESCRIPTION
ØØØØ	Ø	Chip directional register.
ØØØ1	1	Chip I/O: memory paging and tape control.
ØØØ3–ØØØ4	3–4	Float-fixed vector.
ØØØ5–ØØØ6	5–6	Fixed float vector.
ØØØ7	7	Search character.
ØØØ8	8	Scan quotes flag.
ØØØ9	9	TAB column save (temporary).
ØØØA	1Ø	Ø = LOAD , 1 = VERIFY.
ØØØB	11	Input buffer pointer / # subscript.
ØØØC	12	Default DIM flag.
ØØØD	13	Type : FF = string , ØØ = numeric.

000E	14	Type : 80 = integer , 00 = floating point.
000F	15	DATA scan/LIST quote/memory flag.
0010	16	Subscript/FNx flag.
0011	17	0 = INPUT , 40 = GET , 98 = READ (Hex).
0012	18	ATN sign/comparison evaluation flag.
0013	19	Current I/O prompt flag.
0014–0015	20–21	Integer value.
0016	22	Pointer : temporary string stack.
0017–0018	23–24	Last temporary string vector.
0019–0021	25–33	Stack for temporary strings.
0022–0025	34–37	Utility pointer area.
0026–002A	38–42	Product area for multiplication.
002B–002C	43–44	Pointer. . .start of BASIC in low/high byte order.
002D–002E	45–46	Pointer. . .start of variables.
002F–0030	47–48	Pointer. . .start of arrays.
0031–0032	49–50	Pointer. . .end of arrays.
0033–0034	51–52	Pointer. . .string storage (moving down).
0035–0036	53–54	Utility string pointer.
0037–0038	55–56	Pointer. . .limit of memory (top of the memory for BASIC).
0039–003A	57–58	Current BASIC line number.
003B–003C	59–60	Previous BASIC line number.
003D–003E	61–62	Pointer. . .BASIC statement for CONT.
003F–0040	63–64	Current DATA line number.
0041–0042	65–66	Current DATA address.
0043–0044	67–68	INPUT vector.
0045–0046	69–70	Current variable name.
0047–0048	71–72	Current variable address.
0049–004A	73–74	Variable pointer for FOR NEXT.

004B–004C	75–76	Y save; op-save; BASIC pointer save.
004D	77	Comparison symbol accumulator.
004E–0053	78–83	Miscellaneous work area, pointers, etc.
0054–0056	84–86	Jump vector for functions.
0057–0060	87–96	Miscellaneous numeric work area.
0061	97	Accumulator # 1 :exponent.
0062–0065	98–101	Accumulator # 1 :mantissa.
0066	102	Accumulator # 1 :sign.
0067	103	Series evaluation constant pointer.
0068	104	Accumulator # 1 :hi-order (overflow).
0069–006E	105–110	Accumulator # 2 exponent, mantissa, sign (as 97–102 dec).
006F	111	Sign comparison, accumulator # 1 versus accumulator # 2.
0070	112	Accumulator # 1 lo-order (rounding).
0071–0072	113–114	Cassette buffer length/series pointer.
0073–008A	115–138	CHRGET subroutine; get BASIC character.
007A–007B	122–123	BASIC pointer (within CHRGET subroutine).
008B–008F	139–143	RND function seed value.
0090	144	Status word ST.
0091	145	Keyswitch PIA :STOP and RVS flags.
0092	146	Timing constant for tape.
0093	147	Load = 0 , Verify = 1.
0094	148	Serial output :deferred character flag.

ØØ95	149	Serial deferred character.
ØØ96	15Ø	Tape 'End of Tape' marker received.
ØØ97	151	Register save.
ØØ98	152	How many files open.
ØØ99	153	Input device; normally zero.
ØØ9A	154	Output CMD device; normally three.
ØØ9B	155	Tape character parity.
ØØ9C	156	Byte-received flag.
ØØ9D	157	Direct = $ 8Ø/ RUN = Ø. Output control.
ØØ9E	158	To pass one error log/character buffer.
ØØ9F	159	To pass two error log corrected.
ØØAØ–ØØA2	16Ø–162	Jiffy clock (high, med, and low.) Used to set and run TI and TI$:.
ØØA3	163	Serial bit count.
ØØA4	164	Cycle count.
ØØA5	165	Countdown, tape write bit count.
ØØA6	166	Tape buffer pointer.
ØØA7	167	Tape write leader count/Rd pass/inbit.
ØØA8	168	Tape write new byte/Rd error/inbit count.
ØØA9	169	Tape start bit/Rd bit/st bit.
ØØAA	17Ø	Tape scan; cnt; ld; end byte assy.
ØØAB	171	Write lead length/read checksum/parity.
ØØAC–ØØAD	172–173	Pointer: tape buffer, scrolling.
ØØAE–ØØAF	174–175	Tape ends adds/'End of program' marker.
ØØBØ–ØØB1	176–177	Tape timing constants.
ØØB2–ØØB3	178–179	Pointer: start of tape buffer.

ØØB4	18Ø	1 = tape timer enabled ; bit count.
ØØB5	181	Tape 'End of Tape'/RS232 next bit to send.
ØØB6	182	Read character error/outbyte buffer.
ØØB7	183	Number of characters in file name.
ØØB8	184	Current logical file.
ØØB9	185	Current secondary address.
ØØBA	186	Current device.
ØØBB–ØØBC	187–188	Pointer to file name.
ØØBD	189	Write shift word/read input character.
ØØBE	19Ø	Number of blocks remaining to write/read.
ØØBF	191	Serial word buffer.
ØØCØ	192	Tape motor interlock.
ØØC1–ØØC2	193–194	I/O start address.
ØØC3–ØØC4	195–196	KERNAL setup pointer.
ØØC5	197	Current key pressed (64 if no key).
ØØC6	198	Number of characters in keyboard buffer.
ØØC7	199	Screen reverse flag.
ØØC8	2ØØ	End of line for input pointer.
ØØC9–ØØCA	2Ø1–2Ø2	Input cursor log (row, column).
ØØCB	2Ø3	Same as 197.
ØØCC	2Ø4	Ø = flash cursor.
ØØCD	2Ø5	Cursor timing countdown.
ØØCE	2Ø6	Character under cursor.
ØØCF	2Ø7	Cursor in blink phase.
ØØDØ	2Ø8	Input from screen/from keyboard.
ØØD1–ØØD2	2Ø9–21Ø	Pointer to screen line.
ØØD3	211	Position of cursor on above line.

00D4	212	0 = direct cursor, else programmed.
00D5	213	Current screen line length.
00D6	214	Row where cursor lives.
00D7	215	Last INKEY/checksum/buffer.
00D8	216	Number of INSERTs outstanding.
00D9–00F2	217–242	Screen line link table.
00F3–00F4	243–244	Screen colour pointer.
00F5–00F6	245–246	Keyboard pointer.
00F7–00F8	247–248	RS232 Rev pointer.
00F9–00FA	249–250	RS232 Tx pointer.
00FB–00FE	251–256	Free zero page memory.
00FF–010A	256–266	Floating to ASCII work area.
0100–013E	256–318	Tape error log.
0100–01FF	256–511	Processor stack area (the stack).
0200–0258	512–600	BASIC input buffer.
0259–0262	601–610	Logical file table.
0263–026C	611–620	Device number table.
026D–0276	621–630	Secondary address table.
0277–0280	631–640	Keyboard buffer.
0281–0282	641–642	Start of BASIC memory.
0283–0284	643–644	Top of BASIC memory.
0285	645	Serial bus timeout flag.
0286	646	Current colour code (characters).
0287	647	Colour under cursor.
0288	648	Screen memory page.
0289	649	Maximum size of keyboard buffer.
028A	650	Keyboard repeat (128 = repeat all keys, 127 = none).
028B	651	Repeat speed counter.
028C	652	Repeat delay counter.
028D	653	Keyboard shift/CTRL flag (PEEKed to find if SHIFT, etc, pressed.

028E	654	Last shift pattern.
028F–0290	655–656	Keyboard table setup pointer.
0291	657	Keyboar shift mode.
0292	658	0 = scroll enable.
0293	659	RS232 control register.
0294	660	RS232 command register.
0295–0296	661–662	Bit timing.
0297	663	RS232 status.
0298	664	Number of bits to send.
0299–029A	665–666	RS232 speed/code.
029B	667	RS232 receive pointer.
029C	668	RS232 input pointer.
029D	669	RS232 transmit pointer.
029E	670	RS232 output pointer.
029F–02A0	671–672	IRQ save during tape I/O.
02A1	673	CIA 2 (NMI) interrupt control.
02A2	674	CIA 1 timer A control log.
02A3	675	CIA 1 interrupt log.
02A4	676	CIA 1 timer A enabled flag.
02A5	677	Screen row marker.
02C0–02FE	704–766	Block 11 for sprite data.
0300–0301	768–769	Error message link.
0302–0303	770–771	BASIC warm start link.
0304–0305	772–773	Crunch BASIC tokens link.
0306–0307	774–775	PRINT tokens link.
0308–0309	776–777	Start new BASIC code link.
030A–030B	778–779	Get arithmetic element link.
030C	780	System temporary storage of accumulator.
030D	781	System temporary storage of the X register.
030E	782	System temporary storage of the Y register.
030F	783	System status register storage.
0310–0311	784–785	USR function jump (normal B248).
0314–0315	788–789	Hardware interrupt vector (IRQ) (normal EA31).

0316–0317	790–791	Break interrupt vector (normal FE66).
0318–0319	792–793	NMI interrupt vector (normal FE47).
031A–031B	794–795	OPEN vector (normal F34A).
031C–031D	796–797	CLOSE vector (normal F291).
031E–031F	798–799	Set INPUT vector (normal F20E).
0320–0321	800–801	Set OUTPUT vector (normal F250).
0322–0323	802–803	Restore I/O vector (normal F333).
0324–0325	804–805	INPUT vector (normal F157).
0326–0327	806–807	Output vector (normal F1CA).
0328–0329	808–809	Test STOP vector (normal F6ED).
032A–032B	810–811	GET vector (normal F13E).
032C–032D	812–813	Abort I/O vector (normal F32F).
032E–032F	814–815	Warm start vector (normal FE66).
0330–0331	816–817	LOAD link (normal F4A5).
0332–0333	818–819	SAVE link (normal F5ED).
033C–03FB	828–1019	Cassette buffer.
0340–037E	832–894	Block 13 for sprite data.
0380–03BE	896–958	Block 14 for sprite data.
03C0–03FE	960–1022	Block 15 for sprite data.
0400–07E7	1024–2023	Screen memory.
07F8–7FFF	2040–2047	Pointers for sprite data.
0800–9FFF	2048–40959	BASIC RAM memory.
A000–BFFF	40960–49151	BASIC ROM.
A000–BFFF	40960–49151	Paged RAM (behind ROM).
8000–9FFF	32768–40959	Alternative ROM plug in area.
C000–CFFF	49152–53247	RAM (useable in machine code only).
D000–D02E	53248–53294	Video Chip.
D000	53248	Sprite 0 X component.
D001	53249	Sprite 0 Y component.

DØØ2	5325Ø	Sprite 1 X component.
DØØ3	53251	Sprite 1 Y component.
DØØ4	53252	Sprite 2 X component.
DØØ5	53253	Sprite 2 Y component.
DØØ6	53254	Sprite 3 X component.
DØØ7	53255	Sprite 3 Y component.
DØØ8	53256	Sprite 4 X component.
DØØ9	53257	Sprite 4 Y component.
DØØA	53258	Sprite 5 X component.
DØØB	53259	Sprite 5 Y component.
DØØC	5326Ø	Sprite 6 X component.
DØØD	53261	Sprite 6 Y component.
DØØE	53262	Sprite 7 X component.
DØØF	53263	Sprite 7 Y component.
DØ1Ø	53264	MSB of X co-ordinate (for last part of screen).
DØ11	53265	Bit mapped mode and vertical pixel scrolling.
DØ12	53266	Raster.
DØ13	53267	Light pen X component.
DØ14	53268	Light pen Y component.
DØ15	53269	Sprite enable (ON/OFF).
DØ16	5327Ø	Set multicolour character mode. x scroll (pixel).
DØ17	53271	Sprite expand Y.
DØ18	53272	Screen character memory (change character set pointer).
DØ19	53273	Interrupt requests.
DØ1A	53274	Interrupt request MASKS (disable sprite interrupts).
DØ1B	53275	Background/sprite priority.
DØ1C	53276	Multicolour sprite mode.
DØ1D	53277	Sprite expand Y.
DØ1E	53278	Sprite/sprite collision.
DØ1F	53279	Sprite/background collision.
DØ2Ø	5328Ø	Border colour.
DØ21	53281	Screen colour.
DØ22	53282	Multicolour 1 for characters

		(bit pair 01).
D023	53283	Multicolour 2 for characters (bit pair 10).
D024	53284	Background 3.
D025	53285	Sprite (all) multicolour 1 (bit pair 01).
D026	53286	Sprite (all) multicolour 2 (bit pair 11).
D027	53287	Sprite 0 foreground colour (in multicolour it is bit pair 10).
D028	53288	Sprite 1 foreground colour (in multicolour it is bit pair 10).
D029	53289	Sprite 2 foreground colour (in multicolour it is bit pair 10).
D02A	53290	Sprite 3 foreground colour (in multicolour it is bit pair 10).
D02B	53291	Sprite 4 foreground colour (in multicolour it is bit pair 10).
D02C	53292	Sprite 5 foreground colour (in multicolour it is bit pair 10).
D02D	53293	Sprite 6 foreground colour (in multicolour it is bit pair 10).
D02E	53294	Sprite 7 foreground colour (in multicolour it is bit pair 10).
D400–D41C	54272–54300	Sound chip (SID).
D400	54272	Voice 1 low frequency.
D401	54273	Voice 1 high frequency.
D402	54274	Low pulse (pulse waveform only) voice 1.
D403	54275	High pulse (pulse waveform only) voice 1.
D404	54276	Waveform voice 1.
D405	54277	Attack/decay voice 1.
D406	54278	Sustain/release voice 1.
D407	54279	Low frequency voice 2.
D408	54280	High frequency voice 2.
D409	54281	Low pulse (pulse waveform only) voice 2.

D40A	54282	High pulse (pulse waveform only) voice 2.	
D40B	54283	Waveform voice 2.	
D40C	54284	Attack/decay voice 2.	
D40D	54285	Sustain/release voice 2.	
D40E	54286	Low frequency voice 3.	
D40F	54287	High frequency voice 3.	
D410	54288	Low pulse (pulse waveform only) voice 3.	
D411	54289	High pulse (pulse waveform only) voice 3.	
D412	54290	Waveform voice 3.	
D413	54291	Attack/decay voice 3.	
D414	54292	Sustain/release voice 3.	
D415	54293	Low cutoff frequency (0–7).	
D416	54294	High cutoff frequency (0–255).	
D417	54295	Resonance (bits 4–7). Filter voice 3 (turn off) bit 2. Filter voice 2 (bit 1). Filter voice 1 (bit 0).	
D418	54296	High pass filter (bit 6). Low pass filter (bit 4). Band pass filter (bit 5). Master volume control (bits 0–3 (0–15)).	
D800–DBFF	55296–56295	Colour (screen) memory.	
D419	54297	Read paddle X	)
D41A	54298	Read paddle Y	) READ ONLY . . .
D41B	54299	Noise (random)	)
D41C	54300	Envelope 3	)
DC00–DC0F	56320–56335	Interface chip 1(IRQ)(CIA 6526). (See block diagrams.)	

DDØØ–DDØF	56576–56591	Interface chip 2(NMI)(CIA 6526). (See block diagrams.)
EØØØ–FFFF	57344–65535	ROM: operating system (KERNAL).
EØØØ–FFFF	57344–65535	RAM(paged in).
FF81–FFF5	654Ø9–65525	Jump table (to KERNAL subroutines).

KERNAL subroutine (User callable)

ACPTR	FFA5	65445	Input byte from serial port.
CHKIN	FFC6	65478	Open channel for input.
CHKOUT	FFC9	65481	Open channel for output.
CHRIN	FFCF	65487	Input character from channel.
CHROUT	FFD2	6549Ø	Output character to channel. (LDA with ASCII code then JSR FFD2 to print a character on the screen in machine code.)
CIOUT	FF18	65448	Output byte to serial channel.
CLALL	FFE7	65511	Close all channels or files.
CLOSE	FFC3	65475	Close specified logical file.
CLRCHN	FFCC	65484	Close all input and output channels.
GETIN	FFE4	65512	Get character from keyboard buffer.
IOBASE	FFF3	65523	Returns base address of I/O devices.
LISTEN	FFB1	65457	Command devices on the serial bus to listen.
LOAD	FFD5	65493	Load RAM from a device.
MEMBOT	FF9C	65436	Read/set the bottom of memory.
MEMTOP	FF89	65433	Read/set the top of memory.
OPEN	FFCØ	65472	Open a logical file.
PLOT	FFFØ	6552Ø	Read/set X,Y cursor position Load X with X co-ordinate

			(1 –40) and Y with Y co-ordinate then JSR FFF0 (or 65520 in BASIC) by POKEIng into 781 and 782 (X and Y register storage).
RDTIM	FFDE	65502	Read real time clock.
READST	FFB7	65463	Read status word.
RESTOR	FF8A	65415	Restore default I/O vectors.
SAVE	FFD8	65496	Save RAM to a device.
SCNKEY	FF9F	65439	Scan keyboard.
SCREEN	FFED	65517	Returns X,Y organisation of screen.
SECOND	FF93	65427	Send secondary address after LISTEN.
SETLFS	FFBA	65466	Set logical first and second addresses.
SETMSG	FF90	65424	Control KERNAL messages.
SETNAM	FFBD	65469	Set file name.
SETTIM	FFDB	65499	Set real time clock.
SETTMO	FFA2	65442	Set timeout on serial bus.
STOP	FFE1	65505	Scan stop key.
TALK	FFB4	65460	Command serial bus device to TALK.
TKSA	FF96	65430	Second secondary address after TALK.
UDTIM	FFEA	65514	Increment real time clock.
UNLSN	FFAB	65454	Command serial bus to UNLISTEN.
UNTALK	FFAB	65451	Command serial bus to UNTALK.
VECTOR	FF84	65412	Read/set vectored I/O.

PROCERROR SOUND, CIA 1 AND CIA 2 BLOCK DIAGRAMS

Following are charts which contain the memory addresses and functions of the sound chip and the two CAIs.

SID (6581) Commodore 64

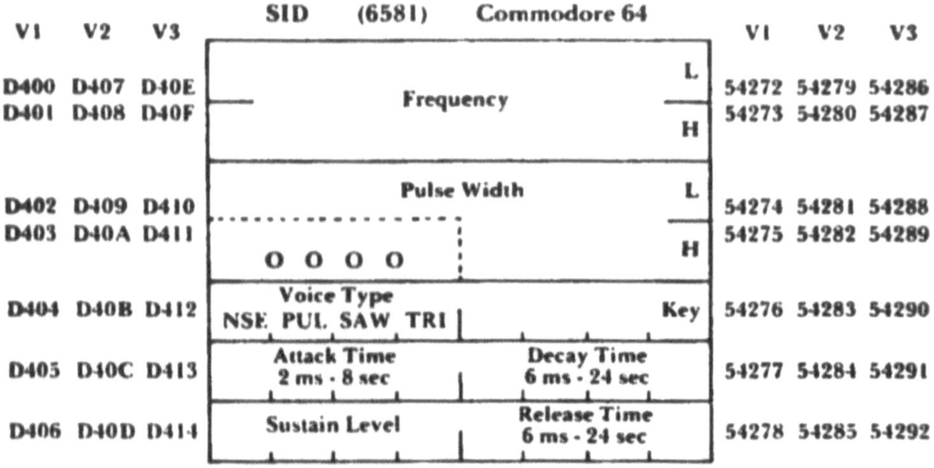

Voices
(Write Only)

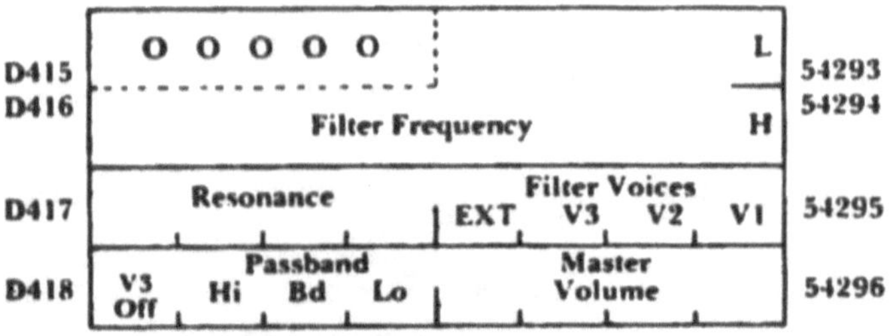

Filter & Volume
(Write Only)

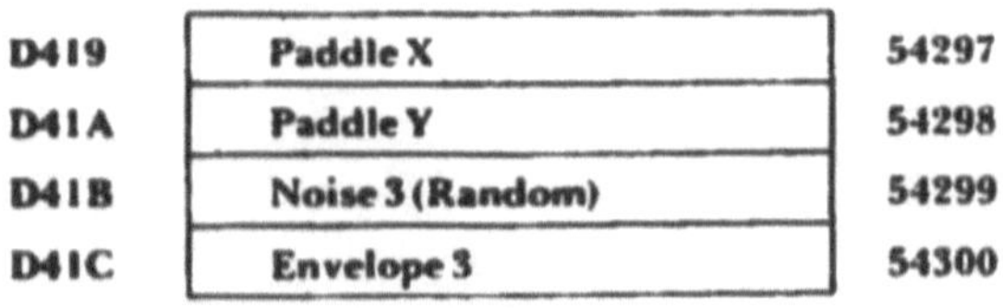

Sense
(Read Only)

Special voice features (TEST, RING MOD, SYNC) are omitted from the above diagram.

Processor I/O Port (6510) Commodore 64

$0000	IN	IN	Out	IN	Out	Out	Out	Out	DDR	0
$0001			Tape Motor	Tape Sense	Tape Write	D-ROM Switch	EF.RAM Switch	AB.RAM Switch	PR	1

CIA 1 (IRQ) (6526) Commodore 64

$DC00	Paddle SEL A	Paddle SEL B			Joystick 0 R	Joystick 0 L	Joystick 0 D	Joystick 0 U	PRA	56320
	Keyboard Row Select (Inverted)									
$DC01					Joystick 1				PRB	56321
	Keyboard Column Read									
$DC02	$FF – All Output								DDRA	56322
$DC03	$00 – All Input								DDRB	56323
$DC04	Timer A								TAL	56324
$DC05									TAH	56325
$DC06	Timer B								TBL	56326
$DC07									TBH	56327
~										
$DC0D				Tape Input			Timer B	Interr. A	ICR	56333
$DC0E					One Shot	Out Mode	Time PB6 Out	Timer A Start	CRA	56334
$DC0F					One Shot	Out Mode	Time PB7 Out	Timer B Start	CRB	56335

CIA 2 (NMI) (6526) Commodore 64

Address									Register	Decimal
$DD00	Serial In	Clock In	Serial Out	Clock Out	ATN Out	RS-232 Out			PRA	56576
$DD01	DSR In	CTS In		DCD* In	RI* In	DTR Out	RTS Out	RS-232 In	PRB	56577
	Parallel User Port									
$DD02	IN	IN	Out	Out	Out $3F	Out	Out	Out	DDRA	56578
$DD03	$06 For RS-232								DDRB	56579
$DD04 $DD05	Timer A								TAL TAH	56580 56581
$DD06 $DD07	Timer B								TBL TBH	56582 56583
$DD0D				RS-232 In			Timer B	Timer A	ICR	56589
$DD0E								Timer A Start	CRA	56590
$DD0F								Timer B Start	CRB	56591

*Connected but not used by system.

NOTES

You may also enjoy...

Made in United States
Troutdale, OR
11/03/2026

56844897R00114